Developments in Biochemistry and Molecular Biology of Plants

Developments in Biochemistry and Molecular Biology of Plants

Dr. R.A.S. Tomar

Editor

KOROS PRESS LIMITED

London, UK

Developments in Biochemistry and Molecular Biology of Plants

© 2012

Printed in 2017 for Sale in the Indian Subcontinent

Published by
Koros Press Limited
3 The Pines, Rubery B45 9FF, Rednal,
Birmingham, United Kingdom

Tel.: +44-7826-930152
Email: info@korospress.com
www.korospress.com

ISBN: 978-1-78163-024-2

Editor: Dr. R.A.S. Tomar

Printed in UK

British Library Cataloguing in Publication Data
A CIP record for this book is available from the British Library

10 9 8 7 6 5 4 3 2 1

Exclusively distributed by CBS Publishers & Distributors Pvt. Ltd.
Sales & Distribution Rights only for India, Pakistan, Bangladesh, Sri Lanka, Nepal and Bhutan.This book is not to be sold outside these territories.

Contents

exotransferase, EC 2.7.7.31) · Thermostable DNA Polymerases with Proofreading Activity · Reverse Transcriptase (RNA-dependent DNA Polymerase EC 2.7.7.49) · Alkaline Phosphatase (EC 3.1.3.1) · Deoxyribonucleases · Ribonucleases · DNA/RNA Nucleases · Methylases · Other Enzymes

Preface

The cell is one of the most basic units of life. There are millions of different types of cells. There are cells that are organisms onto themselves, such as microscopic amoeba and bacteria cells. And there are cells that only function when part of a larger organism, such as the cells that make up your body. The cell is the smallest unit of life in our bodies. In the body, there are brain cells, skin cells, liver cells, stomach cells, and the list goes on. All of these cells have unique functions and features. And all have some recognizable similarities. All cells have a 'skin', called the plasma membrane, protecting it from the outside environment. The cell membrane regulates the movement of water, nutrients and wastes into and out of the cell. Inside of the cell membrane are the working parts of the cell. At the centre of the cell is the cell nucleus. The cell nucleus contains the cell's DNA, the genetic code that coordinates protein synthesis. In addition to the nucleus, there are many organelles inside of the cell - small structures that help carry out the day-to-day operations of the cell. One important cellular organelle is the ribosome.

Ribosomes participate in protein synthesis. The transcription phase of protein synthesis takes places in the cell nucleus. After this step is complete, the mRNA leaves the nucleus and travels to the cell's ribosomes, where translation occurs. Another important cellular organelle is the mitochondrion. Mitochondria (many mitochondrion) are often referred to as the power plants of the cell because many of the reactions that produce energy take place in mitochondria. Also important in the life of a cell are the lysosomes. Lysosomes are organelles that contain enzymes that aid in the digestion of nutrient molecules and other materials. Below is a labelled diagram of a cell to help you identify some of these structures. There are many different types of cells. One major difference in cells occurs between plant cells and animal cells. While both plant and animal cells contain the structures discussed above, plant cells have some additional specialized structures. Many animals have skeletons to give their body structure and support. Plants do not have a skeleton for support and yet plants don't just flop over in a big spongy mess. This is because of a unique cellular structure called the cell wall. The cell wall is a rigid structure outside of the cell membrane composed mainly of the polysaccharide cellulose. As pictured at left, the cell wall gives the plant cell a defined shape which helps support individual parts of plants. In addition to the cell wall, plant cells contain an organelle called the chloroplast.

The book is a meticulously organized and useful both for teaching and for reference. It is intended to serve plant biology and related disciplines, ranging from molecular biology and biotechnology to biochemistry.

—Editor

1

Tools and Techniques in Cell Biology

Hands on Training

Biology is a hands on science, and biology students are usually required to spend some time in the laboratory. The type of activities a student will perform vary depending on the exact field they are in. Plant geneticist, for example, sometimes spend time out in fields gathering plants, while molecular biologists may use complex equipment such as DNA sequencing machines. Even so, there are some basic techniques that all beginning lab students should understand.

Basic Skills

- Laboratory safety and understanding warning signs
- Preparation of simple solutions
- Sampling and preparation of samples
- Dissection of plants and animals
- How to clean glassware
- Use of the dissecting scope
- Use of the light microscope
- Making accurate drawings of organisms.

Research Labs

Research laboratories often require more advanced skills such as the following:

- Agarose gel Electrophoresis
- Use of a pH meter

- Use of a spectrophotometer
- Knowledge of distillation techniques
- Colony counting on agar plates
- Cell counting on slides containing grids
- Making a blood smear.

Advanced

Specialized work places will require other skills. Some skills suggested by subject.

Anatomy, Botany, General Biology

- Use of a microtome to make slides
- staining techniques.

Medical

- Cell and tissue culture
- Patient sample handling techniques
- Making an accurate white blood cell count
- Antibody testing methods such as ELISA.

Microbiology

- Use of micropipetter
- Knowledge of Southern and related blotting techniques
- Molecular cloning
- Use of restriction enzymes
- Use of PCR (Polymerase Chain Reaction) machines
- Replica plating
- Growth of competent cells.

Although books exist to teach these techniques, experienced lab workers have special tricks and techniques that they do not write down. That is why it is best to learn by studying with an experienced lab worker.

Mitochondria

In cell biology, a mitochondrion (plural mitochondria) is a membrane-enclosed organelle found in most eukaryotic cells. These organelles range from 0.5 to 10 micrometers (ìm) in diameter. Mitochondria are sometimes described as "cellular power plants" because they generate most of the cell's supply of adenosine triphosphate (ATP), used as a source of chemical energy. In addition to supplying cellular energy, mitochondria are involved in a range of other processes, such as signalling, cellular differentiation, cell death, as well as the control of the cell cycle and cell growth. Mitochondria have been implicated in several human diseases, including mitochondrial disorders and cardiac dysfunction, and may play a role in the aging process.

Several characteristics make mitochondria unique. The number of mitochondria in a cell varies widely by organism and tissue type. Many cells have only a single mitochondrion, whereas others can contain several thousand mitochondria. The organelle is composed of compartments that carry out specialized functions. These compartments or regions include the outer membrane, the intermembrane space, the inner membrane, and the cristae and matrix. Mitochondrial proteins vary depending on the tissue and the species. In humans, 615 distinct types of proteins have been identified from cardiac mitochondria; whereas in Murinae (rats), 940 proteins encoded by distinct genes have been reported. The mitochondrial proteome is thought to be dynamically regulated. Although most of a cell's DNA is contained in the cell nucleus, the mitochondrion has its own independent genome. Further, its DNA shows substantial similarity to bacterial genomes.

Structure

A mitochondrion contains outer and inner membranes composed of phospholipid bilayers and proteins. The two membranes, however, have different properties. Because of this double-membraned organization, there are five distinct compartments within the mitochondrion. There is the outer mitochondrial membrane, the intermembrane space (the space between the outer and inner membranes), the inner mitochondrial membrane, the crista space (formed by infoldings of the inner membrane), and the matrix (space within the inner membrane).

Outer Membrane

The outer mitochondrial membrane, which encloses the entire organelle, has a protein-to-phospholipid ratio similar to that of the eukaryotic plasma membrane (about 1:1 by weight). It contains large numbers of integral proteins called *porins*. These porins form channels that allow molecules 5000 Daltons or less in molecular weight to freely diffuse from one side of the membrane to the other. Larger proteins can enter the mitochondrion if a signalling sequence at their N-terminus binds to a large multisubunit protein called translocase of the outer membrane, which then actively moves them across the membrane.

Disruption of the outer membrane permits proteins in the intermembrane space to leak into the cytosol, leading to certain cell death. The mitochondrial outer membrane can associate with the endoplasmic reticulum (ER) membrane, in a structure called MAM (mitochondria-associated ER-membrane). This is important in ER-mitochondria calcium signalling and involved in the transfer of lipids between the ER and mitochondria.

Intermembrane Space

The intermembrane space is the space between the outer membrane and the inner membrane. Because the outer membrane is freely permeable to small molecules, the concentrations of small molecules such as ions and sugars in the intermembrane space is the same as the cytosol. However, large proteins must have a specific signalling sequence to be transported across the outer membrane, the protein composition of this space is different from the protein composition of the cytosol. One protein that is localized to the intermembrane space in this way is cytochrome c.

Inner Membrane

The inner mitochondrial membrane contains proteins with five types of functions:

1. Those that perform the redox reactions of oxidative phosphorylation
2. ATP synthase, which generates ATP in the matrix
3. Specific transport proteins that regulate metabolite passage into and out of the matrix
4. Protein import machinery.
5. Mitochondria fusion and fission protein.

It contains more than 150 different polypeptides, and has a very high protein-to-phospholipid ratio (more than 3:1 by weight, which is about 1 protein for 15 phospholipids). The inner membrane is home to around 1/5 of the total protein in a mitochondrion. In addition, the inner membrane is rich in an unusual phospholipid, cardiolipin. This phospholipid was originally discovered in beef hearts in 1942, and is usually characteristic of mitochondrial and bacterial plasma membranes. Cardiolipin contains four fatty acids rather than two and may help to make the inner membrane impermeable.

Unlike the outer membrane, the inner membrane does not contain porins and is highly impermeable to all molecules. Almost all ions and molecules require special membrane transporters to enter or exit the matrix. Proteins are ferried into the matrix via the translocase of the inner membrane (TIM) complex or via Oxa1. In addition, there is a membrane potential across the inner membrane formed by the action of the enzymes of the electron transport chain.

Cristae

The inner mitochondrial membrane is compartmentalized into numerous cristae, which expand the surface area of the inner mitochondrial membrane, enhancing its ability to produce ATP. For typical liver mitochondria the area of the inner membrane is about five times greater than the outer membrane. This ratio is variable and mitochondria from cells that have a greater demand for ATP, such as muscle cells, contain even more cristae. These folds are studded with small round bodies known as F_1 particles or oxysomes. These are not simple random folds but rather invaginations of the inner membrane, which can affect overall chemiosmotic function.

The Matrix

The matrix is the space enclosed by the inner membrane. It contains about 2/3 of the total protein in a mitochondrion. The matrix is important in the production of ATP with the aid of the ATP synthase contained in the inner membrane. The matrix contains a highly-concentrated mixture of hundreds of enzymes, special mitochondrial ribosomes, tRNA, and several copies of the mitochondrial DNA genome. Of the enzymes, the major functions include oxidation of pyruvate and fatty acids, and the citric acid cycle.

Mitochondria have their own genetic material, and the machinery to manufacture their own RNAs and proteins. A published human mitochondrial DNA sequence revealed 16,569 base pairs encoding 37 total genes: 22 tRNA, 2 rRNA, and 13 peptide genes. The 13 mitochondrial peptides in humans are integrated into the inner mitochondrial membrane, along with proteins encoded by genes that reside in the host cell's nucleus.

Organization and Distribution

Mitochondria are found in nearly all eukaryotes. They vary in number and location according to cell type. A single mitochondrion is often found in unicellular organisms. Conversely, numerous mitochondria are found in human liver cells, with about 1000–2000 mitochondria per cell making up 1/5th of the cell volume. The mitochondria can be found nestled between myofibrils of muscle or wrapped around the sperm flagellum. Often they form a complex 3D branching network inside the cell with the cytoskeleton. The association with the cytoskeleton determines mitochondrial shape, which can affect the function as well. Recent evidence suggests vimentin, one of the components of the cytoskeleton, is critical to the association with the cytoskeleton.

Function

The most prominent roles of mitochondria are to produce ATP (i.e., phosphorylation of ADP) through respiration, and to regulate cellular metabolism. The central set of reactions involved in ATP production are collectively known as the citric acid cycle, or the Krebs Cycle. However, the mitochondrion has many other functions in addition to the production of ATP.

Energy Conversion

A dominant role for the mitochondria is the production of ATP, as reflected by the large number of proteins in the inner membrane for this task. This is done by oxidizing the major products of glucose, pyruvate, and NADH, which are produced in the cytosol.

This process of cellular respiration, also known as aerobic respiration, is dependent on the presence of oxygen. When oxygen is limited, the glycolytic products will be metabolized by anaerobic respiration, a process that is independent of the mitochondria. The production of ATP from glucose has an approximately 13-fold higher yield during aerobic respiration compared to anaerobic respiration. Recently it has been shown that plant mitochondria can produce a limited amount of ATP without oxygen by using the alternate substrate nitrite.

Pyruvate: The Citric Acid Cycle

Each pyruvate molecule produced by glycolysis is actively transported across the inner mitochondrial membrane, and into the matrix where it is oxidized and combined with coenzyme A to form CO_2, acetyl-CoA, and NADH.

The acetyl-CoA is the primary substrate to enter the *citric acid cycle*, also known as the *tricarboxylic acid (TCA) cycle* or *Krebs cycle*. The enzymes of the citric acid cycle are located in the mitochondrial matrix, with the exception of succinate dehydrogenase, which is bound to the inner mitochondrial membrane as part of Complex II. The citric acid cycle oxidizes the acetyl-CoA to carbon dioxide, and, in the process, produces reduced cofactors (three molecules of NADH and one molecule of $FADH_2$) that are a source of electrons for the *electron transport chain*, and a molecule of GTP (that is readily converted to an ATP).

NADH and FADH$_2$: The Electron Transport Chain

The redox energy from NADH and $FADH_2$ is transferred to oxygen (O_2) in several steps via the electron transport chain. These energy-rich molecules are produced within the matrix via the citric acid cycle but are also produced in the cytoplasm by glycolysis. Reducing equivalents from the cytoplasm can be imported via the malate-aspartate shuttle system of antiporter proteins or feed into the electron transport chain using a glycerol phosphate shuttle. Protein complexes in the inner membrane (NADH dehydrogenase, cytochrome c reductase, and cytochrome c oxidase) perform the transfer and the incremental release of energy is used to pump protons (H^+) into the intermembrane space. This process is efficient, but a small percentage of electrons may prematurely reduce oxygen, forming reactive oxygen species such as superoxide. This can cause oxidative stress in the mitochondria and may contribute to the decline in mitochondrial function associated with the aging process.

As the proton concentration increases in the intermembrane space, a strong electrochemical gradient is established across the inner membrane. The protons can return to the matrix through the ATP synthase complex, and their potential energy is used to synthesize ATP from ADP and inorganic phosphate (P_i). This process is called chemiosmosis, and was first described by Peter Mitchell who was awarded the 1978 Nobel Prize in Chemistry for his work. Later, part of the 1997 Nobel Prize in Chemistry was awarded to Paul D. Boyer and John E. Walker for their clarification of the working mechanism of ATP synthase.

Heat Production

Under certain conditions, protons can re-enter the mitochondrial matrix without contributing to ATP synthesis. This process is known as *proton leak* or *mitochondrial uncoupling* and is due to the facilitated diffusion of protons into the matrix. The process results in the unharnessed potential energy of the proton electrochemical gradient being released as heat. The process is mediated by a proton channel called thermogenin, or UCP1. Thermogenin is a 33kDa protein first discovered in 1973. Thermogenin is primarily found in brown adipose tissue, or brown fat, and is responsible for non-shivering thermogenesis. Brown adipose tissue is found in mammals, and is at its highest levels in early life and in hibernating animals. In humans, brown adipose tissue is present at birth and decreases with age.

Storage of Calcium Ions

The concentrations of free calcium in the cell can regulate an array of reactions and is important for signal transduction in the cell. Mitochondria can transiently store calcium, a contributing process for the cell's homeostasis of calcium. In fact, their ability to rapidly take in calcium for later release makes them very good "cytosolic buffers" for calcium. The endoplasmic reticulum (ER) is the most significant storage site of calcium, and there is a significant interplay between the mitochondrion and ER with regard to calcium. The calcium is taken up into the matrix by a calcium uniporter on the inner mitochondrial membrane. It is primarily driven by the mitochondrial membrane potential. Release of this calcium back into the cell's interior can occur via a sodium-calcium exchange protein or via "calcium-induced-calcium-release" pathways. This can initiate calcium spikes or calcium waves with large changes in the membrane potential. These can activate a series of second messenger system proteins that can coordinate processes such as neurotransmitter release in nerve cells and release of hormones in endocrine cells.

Additional Functions

Mitochondria play a central role in many other metabolic tasks, such as:

- Regulation of the membrane potential
- Apoptosis-programmed cell death
- Calcium signalling (including calcium-evoked apoptosis)
- Cellular proliferation regulation
- Regulation of cellular metabolism
- Certain heme synthesis reactions.
- Steroid synthesis.

Some mitochondrial functions are performed only in specific types of cells. For example, mitochondria in liver cells contain enzymes that allow them to detoxify ammonia, a waste product of protein metabolism. A mutation in the genes regulating any of these functions can result in mitochondrial diseases.

Origin

Mitochondria have many features in common with prokaryotes. As a result, they are believed to be originally derived from endosymbiotic prokaryotes. A mitochondrion contains DNA, which is organized as several copies of a single, circular chromosome. This mitochondrial chromosome contains genes for redox proteins such as those of the respiratory chain. The CoRR Hypothesis proposes that this Co-location is required for Redox Regulation.

The mitochondrial genome codes for some RNAs of ribosomes, and the twenty-two tRNAs necessary for the translation of messenger RNAs into protein. The circular structure is also found in prokaryotes, and the similarity is extended by the fact that mitochondrial DNA is organized with a variant genetic code similar to that of Proteobacteria. This suggests that their ancestor, the so-called proto-mitochondrion, was a member of the Proteobacteria. In particular, the proto-mitochondrion was probably related to the rickettsia. However, the exact relationship of the ancestor of mitochondria to the alpha-proteobacteria and whether the mitochondria was formed at the same time or after the nucleus, remains controversial.

The ribosomes coded for by the mitochondrial DNA are similar to those from bacteria in size and structure. They closely resemble the bacterial 70S ribosome and not the 80S cytoplasmic ribosomes, which are coded for by nuclear DNA.

The endosymbiotic relationship of mitochondria with their host cells was popularized by Lynn Margulis. The endosymbiotic hypothesis suggests that mitochondria descended from bacteria that somehow survived endocytosis by another cell, and became incorporated into the cytoplasm. The ability of these bacteria to conduct respiration in host cells that had relied on glycolysis and fermentation would have provided a considerable evolutionary advantage. In a similar manner, host cells with symbiotic bacteria capable of photosynthesis would have had an advantage. The incorporation of symbiotes would have increased the number of environments in which the cells could survive. This symbiotic relationship probably developed 1.7-2 billion years ago.

A few groups of unicellular eukaryotes lack mitochondria: the microsporidians, metamonads, and archamoebae. These groups appear as the most primitive eukaryotes on phylogenetic trees constructed using rRNA information, which once suggested that they appeared before the origin of mitochondria. However, this is now known to be an artifact of long-branch attraction—they are derived groups and retain genes or organelles derived from mitochondria (e.g., mitosomes and hydrogenosomes).

Genome

The human mitochondrial genome is a circular DNA molecule of about 16 kilobases. It encodes 37 genes: 13 for subunits of respiratory complexes I, III, IV and V, 22 for mitochondrial tRNA (for the 20 standard amino acids, plus an extra gene for leucine and serine), and 2 for rRNA. One mitochondrion can contain two to ten copies of its DNA.

As in prokaryotes, there is a very high proportion of coding DNA and an absence of repeats. Mitochondrial genes are transcribed as multigenic transcripts, which are cleaved and polyadenylated to yield mature mRNAs. Not all proteins necessary for mitochondrial function are encoded by the mitochondrial genome; most are coded by genes in the cell nucleus and the corresponding proteins are imported into the mitochondrion. The exact number of genes encoded by the nucleus and the mitochondrial genome differs between species. In general, mitochondrial genomes are circular, although exceptions have been reported. In general, mitochondrial DNA lacks introns, as is the case in the human mitochondrial genome; however, introns have been observed in some eukaryotic mitochondrial DNA, such as that of yeast and protists, including *Dictyostelium discoideum*.

In animals the mitochondrial genome is typically a single circular chromosome that is approximately 16-kb long and has 37 genes. The genes while highly conserved may vary in location. Curiously this pattern is not found in the human body louse (*Pediculus humanus*). Instead this mitochondrial genome is arranged in 18 minicircular chromosomes each of which is 3–4 kb long and has one to three genes. This pattern is also found in other sucking lice but not in chewing lice. Recombination has been shown to occur between the minichromosomes. The reason for this difference is not known.

While slight variations on the standard code had been predicted earlier, none was discovered until 1979, when researchers studying human mitochondrial genes determined that they used an alternative code. Many slight variants have been discovered since, including various alternative mitochondrial codes. Further, the AUA, AUC, and AUU codons are all allowable start codons.

Exceptions to the universal genetic code (UGC) in mitochondria

Organism	Codon	Standard	Novel
Mammalian	AGA, AGG	Arginine	Stop codon
	AUA	Isoleucine	Methionine
	UGA	Stop codon	Tryptophan
Invertebrates	AGA, AGG	Arginine	Serine
	AUA	Isoleucine	Methionine
	UGA	Stop codon	Tryptophan
Yeast	AUA	Isoleucine	Methionine
	UGA	Stop codon	Tryptophan
	CUA	Leucine	Threonine

Some of these differences should be regarded as pseudo-changes in the genetic code due to the phenomenon of RNA editing, which is common in mitochondria. In higher plants, it was thought that CGG encoded for tryptophan and not arginine; however, the codon in the processed RNA was discovered to be the UGG codon, consistent with the universal genetic code for tryptophan. Of note, the arthropod mitochondrial genetic code has undergone parallel evolution within a phylum, with some organisms uniquely translating AGG to lysine.

Mitochondrial genomes have far fewer genes than the bacteria from which they are thought to be descended. Although some have been lost altogether, many have been transferred to the nucleus, such as the respiratory complex II protein subunits. This is thought to be relatively common over evolutionary time. A few organisms, such as the *Cryptosporidium*, actually have mitochondria that lack any DNA, presumably because all their genes have been lost or transferred. In *Cryptosporidium*, the mitochondria have an altered ATP generation system that renders the parasite resistant to many classical mitochondrial inhibitors such as cyanide, azide, and atovaquone.

Replication and Inheritance

Mitochondria divide by binary fission similar to bacterial cell division; unlike bacteria, however, mitochondria can also fuse with other mitochondria.. The regulation of this division differs between eukaryotes. In many single-celled eukaryotes, their growth and division is linked to the cell cycle. For example, a single mitochondrion may divide synchronously with the nucleus. This division and segregation process must be tightly controlled so that each daughter cell receives at least one mitochondrion. In other eukaryotes (in mammals for example), mitochondria may replicate their DNA and divide mainly in response to the energy needs of the cell, rather than in phase with the cell cycle. When the energy needs of a cell are high, mitochondria grow and divide. When the energy use is low, mitochondria are destroyed or become inactive. In such examples, and in contrast to the situation in many single celled eukaryotes, mitochondria are apparently randomly distributed to the daughter cells during the division of the cytoplasm.

An individual's mitochondrial genes are not inherited by the same mechanism as nuclear genes. At fertilization of an egg cell by a sperm, the egg nucleus and sperm nucleus each contribute equally to the genetic make-up of the zygote nucleus. In contrast, the mitochondria, and therefore the mitochondrial DNA, usually comes from the egg only. The sperm's mitochondria enter the egg but do not contribute genetic information to the embryo. Instead, paternal mitochondria are marked with ubiquitin to select them for later destruction inside the embryo. The egg cell contains relatively few mitochondria, but it is these mitochondria that survive and divide to populate the cells of the adult organism. Mitochondria are, therefore, in most cases inherited down the female line, known as maternal inheritance. This mode is seen in most organisms including all animals. However, mitochondria in some species can sometimes be inherited paternally. This is the norm among certain coniferous plants, although not in pine trees and yew trees. It has been suggested that it occurs at a very low level in humans.

Uniparental inheritance leads to little opportunity for genetic recombination between different lineages of mitochondria, although a single mitochondrion can contain 2–10 copies of its DNA. For this reason, mitochondrial DNA usually is thought to reproduce by binary fission. What recombination does take place maintains genetic integrity rather than maintaining diversity. However, there are studies showing evidence of recombination in mitochondrial DNA. It is clear that the enzymes necessary for recombination are present in mammalian cells. Further, evidence suggests that animal mitochondria can undergo recombination. The data are a bit more controversial in humans, although indirect evidence of recombination exists. If recombination does not occur, the whole mitochondrial DNA sequence represents a single haplotype, which makes it useful for studying the evolutionary history of populations.

Population Genetic Studies

The near-absence of genetic recombination in mitochondrial DNA makes it a useful source of information for scientists involved in population genetics and evolutionary biology. Because all the mitochondrial DNA is inherited as a single unit, or haplotype, the relationships between mitochondrial DNA from different individuals can be represented as a gene tree. Patterns in these gene trees can be used to infer the evolutionary history of populations. The classic example of this is in human evolutionary genetics, where the molecular clock can be used to provide a recent date for mitochondrial Eve. This is often interpreted as strong support for a recent modern human expansion out of Africa. Another human example is the sequencing of mitochondrial DNA from Neanderthal bones. The relatively large evolutionary distance between the mitochondrial DNA sequences of Neanderthals and living humans has been interpreted as evidence for lack of interbreeding between Neanderthals and anatomically-modern humans.

However, mitochondrial DNA reflects the history of only females in a population and so may not represent the history of the population as a whole. This can be partially overcome by the use of paternal genetic sequences, such as the non-recombining region of the Y-chromosome. In a broader sense, only studies that also include nuclear DNA can provide a comprehensive evolutionary history of a population.

Dysfunction and Disease

Mitochondrial Diseases

With their central place in cell metabolism, damage — and subsequent dysfunction — in mitochondria is an important factor in a wide range of human diseases. Mitochondrial disorders often present as neurological disorders, but can manifest as myopathy, diabetes, multiple endocrinopathy, or a variety of other systemic manifestations. Diseases caused by mutation in the mtDNA include Kearns-Sayre syndrome, MELAS syndrome and Leber's hereditary optic neuropathy. In the vast majority of cases, these diseases are transmitted by a female to her children, as the zygote derives its mitochondria and hence its mtDNA from the ovum. Diseases such as Kearns-Sayre syndrome, Pearson's syndrome, and progressive external ophthalmoplegia are thought to be due to large-scale mtDNA rearrangements, whereas other diseases such as MELAS syndrome, Leber's hereditary optic neuropathy, myoclonic epilepsy with ragged red fibres (MERRF), and others are due to point mutations in mtDNA. In other diseases, defects in nuclear genes lead to dysfunction of mitochondrial proteins. This is the case in Friedreich's ataxia, hereditary spastic paraplegia, and Wilson's disease. These diseases are inherited in a dominance relationship, as applies to most other genetic diseases. A variety of disorders can be caused by nuclear mutations of oxidative phosphorylation enzymes, such as coenzyme Q10 deficiency and Barth syndrome. Environmental influences may interact with hereditary predispositions and cause mitochondrial disease. For example, there may be a link between pesticide exposure and the later onset of Parkinson's disease.

Other pathologies with etiology involving mitochondrial dysfunction include schizophrenia, bipolar disorder, dementia, Alzheimer's disease, Parkinson's disease, epilepsy, stroke, cardiovascular disease, retinitis pigmentosa, and diabetes mellitus. A common thread thought to link these seemingly-unrelated conditions is cellular damage causing oxidative stress. How exactly mitochondrial dysfunction fits into the etiology of these pathologies is yet to be elucidated.

Possible Relationships to Aging

Given the role of mitochondria as the cell's powerhouse, there may be some leakage of the high-energy electrons in the respiratory chain to form reactive oxygen species. This can result in significant oxidative stress in the mitochondria with high mutation rates of mitochondrial DNA. A vicious cycle is thought to occur, as oxidative stress leads to mitochondrial DNA mutations, which can lead to enzymatic abnormalities and further oxidative stress. A number of changes occur to mitochondria during the aging process. Tissues from elderly patients show a decrease in enzymatic activity of the proteins of the respiratory chain. Large deletions in the mitochondrial genome can lead to high levels of oxidative stress and neuronal death in Parkinson's disease. Hypothesized links between aging and oxidative stress are not new and were proposed over 50 years ago; however, there is much debate over whether mitochondrial changes are causes of aging or merely characteristics of aging. One notable study in mice demonstrated shortened lifespan but no increase in reactive oxygen species despite increasing mitochondrial DNA mutations, suggesting that mitochondrial DNA mutations can cause lifespan shortening by other mechanisms. As a result, the exact relationships between mitochondria, oxidative stress, and aging have not yet been settled.

Plastids

Plastids are major organelles found in the cells of plants and algae. Plastids are the site of manufacture and storage of important chemical compounds used by the cell. Plastids often contain pigments used in photosynthesis, and the types of pigments present can change or determine the cell's color.

Plastids in Plants

Plastids are responsible for photosynthesis, storage of products like starch and for the synthesis have the ability to differentiate, or redifferentiate, between these and other forms. All plastids are derived from proplastids (formerly "eoplasts", *eo-*: dawn, early), which are present in the meristematic regions of the plant. Proplastids and young chloroplasts commonly divide, but more mature chloroplasts also have this capacity. In plants, plastids may differentiate into several forms, depending upon which function they need to play in the cell. Undifferentiated plastids (*proplastids*) may develop into any of the following plastids:

- Chloroplasts: for photosynthesis; *see also etioplasts, the predecessors of chloroplasts*
- Chromoplasts: for pigment synthesis and storage
- Gerontoplasts: control the dismantling of the photosynthetic apparatus during senescence
- Leucoplasts: for monoterpene synthesis; *leucoplasts sometimes differentiate into more specialized plastids:*
    — Amyloplasts: for starch storage and detecting gravity
    — Elaioplasts: for storing fat
    — Proteinoplasts: for storing and modifying protein.

Each plastid creates multiple copies of the circular 75-250 kilobase plastome. The number of genome copies per plastid is flexible, ranging from more than 1000 in rapidly dividing cells, which generally contain few plastids, to 100 or fewer in mature cells, where plastid divisions has given rise to a large number of plastids. The plastome contains about 100 genes encoding ribosomal and transfer ribonucleic acids (rRNAs and tRNAs) as well as proteins involved in photosynthesis and plastid gene transcription and translation. However, these proteins only represent a small fraction of the total protein set-up necessary to build and maintain the structure and function of a particular type of plastid. Nuclear genes encode the vast majority of plastid proteins, and the expression of plastid genes and nuclear genes is tightly co-regulated to allow proper development of plastids in relation to cell differentiation.

Plastid DNA exists as large protein-DNA complexes associated with the inner envelope membrane and called 'plastid nucleoids'. Each nucleoid particle may contain more than 10 copies of the plastid DNA. The proplastid contains a single nucleoid located in the centre of the plastid. The developing plastid has many nucleoids, localized at the periphery of the plastid, bound to the inner envelope membrane. During the development of proplastids to chloroplasts, and when plastids convert from one type to another, nucleoids change in morphology, size and location within the organelle.

The remodelling of nucleoids is believed to occur by modifications to the composition and abundance of nucleoid proteins. Many plastids, particularly those responsible for photosynthesis, possess numerous internal membrane layers. In plant cells, long thin protuberances called stromules sometimes form and extend from the main plastid body into the cytosol and interconnect several plastids. Proteins, and presumably smaller molecules, can move within stromules. Most cultured cells that are relatively large compared to other plant cells have very long and abundant stromules that extend to the cell periphery.

Plastids in Algae

In algae, the term leucoplast is used for all unpigmented plastids and their function differs from the leucoplasts of plants. Etioplasts, amyloplasts and chromoplasts are plant-specific and do not occur in algae. Plastids in algae and hornworts may also differ from plant plastids in that they contain pyrenoids.

Glaucocystophytic algae contain muroplasts which are similar to chloroplasts except that they have a cell wall that is similar to that of prokaryotes. Rhydophytic algae contain rhydoplasts which are red chloroplasts which allow the algae to photosynthesise to a depth of up to 268m.

Inheritance of Plastids

Most plants inherit the plastids from only one parent. Angiosperms generally inherit plastids from the female gamete, while many gymnosperms inherit plastids from the male pollen. Algae also inherit plastids from only one parent. The plastid DNA of the other parent is thus completely lost.

In normal intraspecific crossings (resulting in normal hybrids of one species), the inheritance of plastid DNA appears to be quite strictly 100% uniparental. In interspecific hybridisations, however, the inheritance of plastids appears to be more erratic. Although plastids inherit mainly maternally in interspecific hybridisations, there are many reports of hybrids of flowering plants that contain plastids of the father. Approximately ~20% of angiosperms, including alfalfa (Medicago), normally show biparental inheritance of plastids.

Origin of Plastids

Plastids are thought to have originated from endosymbiotic cyanobacteria. They developed around 1500 million years ago and allowed eukaryotes to carry out oxygenic photosynthesis. Due to a split-up into three evolutionary lineages, the plastids are named differently: chloroplasts in green algae and plants, rhodoplasts in red algae and cyanelles in the glaucophytes.

The plastids differ by their pigmentation, but also in ultrastructure. The chloroplasts e.g. have lost all phycobilisomes, the light harvesting complexes found in cyanobacteria, red algae and glaucophytes, but only in plants and in closely related green algae contain stroma and grana thylakoids. The glaucocystophycean plastid in contrast to the chloroplasts and the rhodoplasts is still surrounded by the remains of the cyanobacterial cell wall. All these primary plastids are surrounded by two membranes.

Complex plastids start by secondary endosymbiosis, when a eukaryote engulfs a red or green alga and retains the algal plastid, which is typically surrounded by more than two membranes. In some cases these plastids may be reduced in their metabolic and/or photosynthetic capacity. Algae with complex plastids derived by secondary endosymbiosis of a red alga include the heterokonts, haptophytes, cryptomonads, and most dinoflagellates (= rhodoplasts). Those that endosymbiosed a green alga include the euglenids and chlorarachniophytes (= chloroplasts). The Apicomplexa, a phylum of obligate parasitic protozoa including the causative agents of malaria (*Plasmodium* spp.), toxoplasmosis (*Toxoplasma gondii*), and many other human or animal diseases also harbour a complex plastid (although this organelle has been lost in some apicomplexans, such as *Cryptosporidium parvum*, which causes cryptosporidiosis).

The 'apicoplast' is no longer capable of photosynthesis, but is an essential organelle, and a promising target for antiparasitic drug development. Some dinoflagellates and sea slugs, particularly of the genus Elysia, take up algae as food and keep the plastid of the digested alga to profit from the photosynthesis; after a while the plastids are also digested. These captured plastids are known as kleptoplastids.

Endoplasmic Reticulum

The endoplasmic reticulum (ER) is an eukaryotic organelle that forms an interconnected network of tubules, vesicles, and cisternae within cells. Rough endoplasmic reticulums synthesize proteins, while smooth endoplasmic reticulums synthesize lipids and steroids, metabolize carbohydrates and steroids, and regulate calcium concentration, drug detoxification, and attachment of receptors on cell membrane proteins. Sarcoplasmic reticulums solely regulate calcium levels. The lacey membranes of the endoplasmic reticulum were first seen by Keith R. Porter, Albert Claude, and Ernest F. Fullam in 1945.

Structure

The general structure of the endoplasmic reticulum is an extensive membrane network of cisternae (sac-like structures) held together by the cytoskeleton. The phospholipid membrane encloses a space, the cisternal space (or lumen), from the cytosol, which is continuous with the perinuclear space. The functions of the endoplasmic reticulum vary greatly depending on the exact type of endoplasmic reticulum and the type of cell in which it resides. The three varieties are called *rough endoplasmic reticulum, smooth endoplasmic reticulum* and *sarcoplasmic reticulum.*

The quantity of RER and SER in a cell can quickly interchange from one type to the other, depending on changing metabolic needs: one type will undergo numerous changes including new proteins embedded in the membranes in order to transform. Also, massive changes in the protein content can occur without any noticeable structural changes, depending on the enzymatic needs of the cell.

Rough Endoplasmic Reticulum

The surface of the rough endoplasmic reticulum (RER) is studded with protein-manufacturing ribosomes giving it a "rough" appearance (hence its name). However, the ribosomes bound to the RER at any one time are not a stable part of this organelle's structure as ribosomes are constantly being bound and released from the membrane. A ribosome only binds to the ER once it begins to synthesize a protein destined for the secretory pathway. Here, a ribosome in the cytosol begins synthesizing a protein until a signal recognition particle recognizes the pre-piece of 5-15 hydrophobic amino acids preceded by a positively charged amino acid. This signal sequence allows the recognition particle to bind to the ribosome, causing the ribosome to bind to the RER and pass the new protein through the ER membrane. The pre-piece is then cleaved off within the lumen of the ER and the ribosome released back into the cytosol.

The membrane of the RER is continuous with the outer layer of the nuclear envelope. Although there is no continuous membrane between the RER and the Golgi apparatus, membrane-bound vesicles shuttle proteins between these two compartments. Vesicles are surrounded by coating proteins called COPI and COPII. COPII targets vesicles to the golgi and COPI marks them to be brought back to the RER. The RER works in concert with the Golgi complex to target new proteins to their proper destinations. A second method of transport out of the ER are areas called membrane contact sites, where the membranes of the ER and other organelles are held closely together, allowing the transfer of lipids and other small molecules.

The RER is key in multiple functions:

- lysosomal enzymes with a mannose-6-phosphate marker added in the *cis*-Golgi network
- Secreted proteins, either secreted constitutively with no tag, or regulated secretion involving clathrin and paired basic amino acids in the signal peptide.
- integral membrane proteins that stay imbedded in the membrane as vesicles exit and bind to new membranes. Rab proteins are key in targeting the membrane, SNAP and SNARE proteins are key in the fusion event.
- initial glycosylation as assembly continues. This is either N-linked (O-linking occur in the golgi).

- N-linked glycosylation: if the protein is properly folded, glycosyltransferase recognizes the AA sequence NXS or NXT (with the S/T residue phosphorylated) and adds a 14 sugar backbone (2 *N*-acetylglucosamine, 9 branching mannose, and 3 glucose at the end) to the side chain nitrogen of Asn.

Smooth Endoplasmic Reticulum

The smooth endoplasmic reticulum (SER) has functions in several metabolic processes, including synthesis of lipids and steroids, metabolism of carbohydrates, regulation of calcium concentration, drug detoxification, attachment of receptors on cell membrane proteins, and steroid metabolism. It is connected to the nuclear envelope. Smooth endoplasmic reticulum is found in a variety of cell types (both animal and plant) and it serves different functions in each. The Smooth ER also contains the enzyme glucose-6-phosphatase which converts glucose-6-phosphate to glucose, a step in gluconeogenesis. The SER consists of tubules and vesicles that branch forming a network. In some cells there are dilated areas like the sacs of RER. The network of SER allows increased surface area for the action or storage of key enzymes and the products of these enzymes.

Sarcoplasmic Reticulum

The sarcoplasmic reticulum (SR), from the Greek *sarx*, ("flesh"), is a special type of smooth ER found in smooth and striated muscle. The only structural difference between this organelle and the SER is the medley of proteins they have, both bound to their membranes and drifting within the confines of their lumens. This fundamental difference is indicative of their functions: the SER synthesizes molecules while the SR stores and pumps calcium ions. The SR contains large stores of calcium, which it sequesters and then releases when the muscle cell is stimulated. The SR's release of calcium upon electrical stimulation of the cell plays a major role in excitation-contraction coupling.

Functions

The endoplasmic reticulum serves many general functions, including the facilitation of protein folding and the transport of synthesized proteins in sacs called cisternae. Correct folding of newly-made proteins is made possible by several endoplasmic reticulum chaperone proteins, including protein disulfide isomerase (PDI), ERp29, the Hsp70 family member Grp78, calnexin, calreticulin, and the peptidylpropyl isomerase family. Only properly-folded proteins are transported from the rough ER to the Golgi complex.

Transport of Proteins

Secretory proteins, mostly glycoproteins, are moved across the endoplasmic reticulum membrane. Proteins that are transported by the endoplasmic reticulum and from there throughout the cell are marked with an address tag called a signal sequence. The N-terminus (one end) of a polypeptide chain (i.e., a protein) contains a few amino acids that work as an address tag, which are removed when the polypeptide reaches its destination. Proteins that are destined for places outside the endoplasmic reticulum are packed into transport vesicles and moved along the cytoskeleton toward their destination.

The endoplasmic reticulum is also part of a protein sorting pathway. It is, in essence, the transportation system of the eukaryotic cell. The majority of endoplasmic reticulum resident proteins are retained in the endoplasmic reticulum through a retention motif. This motif is composed of four amino acids at the end of the protein sequence. The most common retention sequence is KDEL (*lys-asp-glu-leu*). However, variation on KDEL does occur and other sequences can also give rise to endoplasmic reticulum retention. It is not known if such variation can lead to sub-endoplasmic reticulum localizations. There are three KDEL receptors in mammalian cells, and they have a very high degree of sequence identity. The functional differences between these receptors remain to be established.

Other Functions

- Insertion of proteins into the endoplasmic reticulum membrane: Integral membrane proteins are inserted into the endoplasmic reticulum membrane as they are being synthesized (co-translational translocation). Insertion into the endoplasmic reticulum membrane requires the correct topogenic signal sequences in the protein.

- Glycosylation: Glycosylation involves the attachment of oligosaccharides.

- Disulfide bond formation and rearrangement: Disulfide bonds stabilize the tertiary and quaternary structure of many proteins.

- Drug Metabolism: The smooth ER is the site at which some drugs are modified by microsomal enzymes which include the cytochrome P450 enzymes.

Golgi Complex

Golgi apparatus (also Golgi body or Golgi Complex) is an organelle found in most eukaryotic cells. It was identified in 1897 by the Italian physician Camillo Golgi and was named after him. The primary function of the Golgi apparatus is to process and package macromolecules, such as proteins and lipids, after their synthesis and before they make their way to their destination; it is particularly important in the processing of proteins for secretion. The Golgi apparatus forms a part of the cellular endomembrane system.

Discovery

Due to its fairly large size, the golgi apparatus was one of the first organelles to be discovered and observed in detail. The apparatus was discovered in 1897 by Italian physician Camillo Golgi during an investigation of the nervous system. After first observing it under his microscope, he termed the structure the *internal reticular apparatus.* The structure was then renamed after Golgi not long after the announcement of his discovery in 1898. However, some doubted the discovery at first, arguing that the appearance of the structure was merely an optical illusion created by the observation technique used by Golgi. With the development of modern microscopes in the 20th century, the discovery was confirmed.

Structure

The Golgi is composed of stacks of membrane-bound structures known as cisternae (singular: *cisterna*). An individual stack is sometimes called a dictyosome (from Greek dictyon, net + soma, body), especially in plant cells. A mammalian cell typically contains 40 to 100 stacks. Between four and eight cisternae are usually present in a stack; however, in some protists as many as sixty have been observed. Each cisterna comprises a flattened membrane disk, and carries Golgi enzymes to help or to modify cargo proteins that travel through them. They are found in both plant and animal cells.

The cisternae stack has four functional regions: the cis-Golgi network, medial-Golgi, endo-Golgi, and trans-Golgi network. Vesicles from the endoplasmic reticulum (via the vesicular-tubular clusters) fuse with the network and subsequently progress through the stack to the trans Golgi network, where they are packaged and sent to the required destination. Each region contains different enzymes which selectively modify the contents depending on where they reside. The cisternae also carry structural proteins important for their maintenance as flattened membranes which stack upon each other.

Function

Cells synthesize a large number of different macromolecules. The Golgi apparatus is integral in modifying, sorting, and packaging these macromolecules for cell secretion (exocytosis) or use within the cell. It primarily modifies proteins delivered from the rough endoplasmic reticulum but is also involved in the transport of lipids around the cell, and the creation of lysosomes. In this respect it can be thought of as similar to a post office; it packages and labels items which it then sends to different parts of the cell.

Enzymes within the cisternae are able to modify substances by the addition of carbohydrates (glycosylation) and phosphates (phosphorylation). In order to do so, the Golgi imports substances such as nucleotide sugars from the cytosol. These modifications may also form a signal sequence which determines their final destination. For example, the Golgi apparatus adds a mannose-6-phosphate label to proteins destined for lysosomes.

The Golgi plays an important role in the synthesis of proteoglycans, which are molecules present in the extracellular matrix of animals. It is also a major site of carbohydrate synthesis. This includes the productions of glycosaminoglycans (GAGs), long unbranched polysaccharides which the Golgi then attaches to a protein synthesised in the endoplasmic reticulum to form proteoglycans. Enzymes in the Golgi polymerize several of these GAGs via a xylose link onto the core protein. Another task of the Golgi involves the sulfation of certain molecules passing through its lumen via sulphotranferases that gain their sulphur molecule from a donor called PAPs. This process occurs on the GAGs of proteoglycans as well as on the core protein. The level of sulfation is very important to the proteoglycans' signalling abilities as well as giving the proteoglycan its overall negative charge.

The phosphorylation of molecules requires that ATP is imported into the lumen of the Golgi and then utilised by resident kinases such as casein kinase 1 and casein kinase 2. One molecule that is phosphorylated in the Golgi is Apolipoprotein, which forms a molecule known as VLDL that is a constitute of blood serum. It is thought that the phosphorylation of these molecules is important to help aid in their sorting for secretion into the blood serum.

The Golgi has a putative role in apoptosis, with several Bcl-2 family members localised there, as well as to the mitochondria. A newly characterized protein, GAAP (Golgi anti-apoptotic protein), almost exclusively resides in the Golgi and protects cells from apoptosis by an as-yet undefined mechanism.

Vesicular Transport

The vesicles that leave the rough endoplasmic reticulum are transported to the *cis* face of the Golgi apparatus, where they fuse with the Golgi membrane and empty their contents into the lumen. Once inside they are modified, sorted and shipped towards their final destination. As such, the Golgi apparatus tends to be more prominent and numerous in cells synthesising and secreting many substances: plasma B cells, the antibody-secreting cells of the immune system, have prominent Golgi complexes. Those proteins destined for areas of the cell other than either the endoplasmic reticulum or Golgi apparatus are moved towards the *trans* face, to a complex network of membranes and associated vesicles known as the *trans-Golgi network* (TGN). This area of the Golgi is the point at which proteins are sorted and shipped to their intended destinations by their placement into one of at least three different types of vesicles, depending upon the molecular marker they carry:

Transport Mechanism

The transport mechanism which proteins use to progress through the Golgi apparatus is not yet clear; however a number of hypotheses currently exist. Until recently, the vesicular transport mechanism was favoured but now more evidence is coming to light to support cisternal maturation. The two proposed models may actually work in conjunction with each other, rather than being mutually exclusive. This is sometimes referred to as the *combined* model.

- *Cisternal Maturation Model*: the cisternae of the Golgi apparatus move by being built at the *cis* face and destroyed at the *trans* face. Vesicles from the endoplasmic reticulum fuse with each other to form a cisterna at the *cis* face, consequently this cisterna would appear to move through the Golgi stack when a new cisterna is formed at the *cis* face. This model is supported by the fact that structures larger than the transport vesicles, such as collagen rods, were observed microscopically to progress through the Golgi apparatus. This was initially a popular hypothesis, but lost favour in the 1980s. Recently it has made a comeback, as laboratories at the University of Chicago and the University of Tokyo have been able to use new technology to directly observe Golgi compartments maturing. Additional evidence comes from the fact that COPI vesicles move in the retrograde direction, transporting Endoplasmic Reticulum proteins back to where they belong by recognizing a signal peptide.

- *Vesicular transport model*: Vesicular transport views the Golgi as a very stable organelle, divided into compartments in the cis to trans direction. Membrane bound carriers transport material between the ER and the different compartments of the Golgi. Experimental evidence includes the abundance of small vesicles (known technically as shuttle vesicles) in proximity to the Golgi apparatus. To direct the vesicles, actin filaments connect packaging proteins to the membrane to ensure that they fuse with the correct compartment.

Golgi Apparatus During Mitosis

The Golgi apparatus will break up and disappear following the onset of mitosis, or cellular division. During the telophase of mitosis, the Golgi apparatus reappears; however, it is still uncertain how this occurs.

Golgi Apparatus in Popular Culture

"Golgi Apparatus" is the title of a song by the band Phish in their album Junta. Tom Marshall and guitarist Trey Anastasio wrote this with some friends during grade school and later reworked it as a Phish song.

Lysosomes

Lysosomes are cube shaped organelles that contain enzymes (acid hydrolases) that break up waste materials and cellular debris. They are found in animal cells, while in yeast and plants the same roles are performed by lytic vacuoles. Lysosomes digest excess or worn-out organelles, food particles, and engulfed viruses or bacteria. The membrane around a lysosome allows the digestive enzymes to work at the 4.5 pH they require. Lysosomes fuse with vacuoles and dispense their enzymes into the vacuoles, digesting their contents. They are created by the addition of hydrolytic enzymes to early endosomes from the Golgi apparatus. The name *lysosome* derives from the Greek words *lysis*, which means to separate; and *soma*, which means body. They are frequently nicknamed "suicide-bags" or "suicide-sacs" by cell biologists due to their role in autolysis. Lysosomes were discovered by the Belgian cytologist Christian de Duve in 1949.

The size of lysosomes varies from 0.1–1.2 m. At pH 4.8, the interior of the lysosomes is acidic compared to the slightly alkaline cytosol (pH 7.2). The lysosome maintains this pH differential by pumping protons (H ions) from the cytosol across the membrane via proton pumps and chloride ion channels. The lysosomal membrane protects the cytosol, and therefore the rest of the cell, from the degradative enzymes within the lysosome. The cell is additionally protected from any lysosomal acid hydrolases that leak into the cytosol as these enzymes are pH-sensitive and function less well in the alkaline environment of the cytosol.

Enzymes

Some important enzymes found within lysosomes include:
- Lipase, which digests lipids
- Amylase, which digest Starchs (a kind of carbohydrate)
- Proteases, which digest proteins
- Nucleases, which digest nucleic acids
- phosphoric acid monoesters.

Lysosomal enzymes are synthesized in the cytosol and the endoplasmic reticulum, where they receive a mannose-6-phosphate tag that targets them for the lysosome. Aberrant lysosomal targeting causes inclusion-cell disease, whereby enzymes do not properly reach the lysosome, resulting in accumulation of waste within these organelles.

Functions

Lysosomes are the cell's garbage disposal system. They are used for the digestion of macromolecules from phagocytosis (ingestion of other dying cells or larger extracellular material, like foreign invading microbes), endocytosis (where receptor proteins are recycled from the cell surface), and autophagy (where in old or unneeded organelles or proteins, or microbes that have invaded the cytoplasm are delivered to the lysosome). Autophagy may also lead to autophagic cell death, a form of programmed self-destruction, or autolysis, of the cell, which means that the cell is digesting itself.

Other functions include digesting foreign bacteria (or other forms of waste) that invade a cell and helping repair damage to the plasma membrane by serving as a membrane patch, sealing the wound. In the past, lysosomes were thought to kill cells that were no longer wanted, such as those in the tails of tadpoles or in the web from the fingers of a 3- to 6-month-old fetus. While lysosomes digest some materials in this process, it is actually accomplished through programmed cell death, called apoptosis.

Clinical Relevance

There are a number of lysosomal storage diseases that are caused by the malfunction of the lysosomes or one of their digestive proteins; examples include Tay-Sachs disease and Pompe's disease. These diseases are caused by a defective or missing digestive protein, which leads to the accumulation of substrates within the cell, impairing metabolism. In the broad sense, these can be classified as mucopolysaccharidoses, GM_2 gangliosidoses, lipid storage disorders, glycoproteinoses, mucolipidoses, or leukodystrophies.

2

Cell Reproduction: Cell Growth and Cell Cycle

The process of a cell splitting and becoming two similar cells.

Prokaryotes reproduce in a process called binary fission.

Eukaryotic cells reproduce using either mitosis or meiosis. Mitosis creates two daughter cells with an equal number of chromosomes. Meiosis creates four daughter cells, each with half the number chromosomes as the parent, and is used in sexual reproduction.

Cell Division

Cell division is the process by which a *parent cell* divides into two or more *daughter cells*. Cell division is usually a small segment of a larger cell cycle. This type of cell division in eukaryotes is known as mitosis, and leaves the daughter cell capable of dividing again. The corresponding sort of cell division in prokaryotes is known as binary fission. In another type of cell division present only in eukaryotes, called meiosis, a cell is permanently transformed into a gamete and cannot divide again until fertilization. Right before the parent cell splits, it undergoes DNA replication.

For simple unicellular organisms such as the amoeba, one cell division is equivalent to reproduction— an entire new organism is created. On a larger scale, mitotic cell division can create progeny from multicellular organisms, such as plants that grow from cuttings. Cell division also enables asexually reproducing organisms to develop from the one-celled zygote, which itself was produced by cell division from gametes. And after growth, cell division allows for continual construction and repair of the organism. A human being's body experiences about 10,000 trillion cell divisions in a lifetime. The primary concern of cell division is the maintenance of the original cell's genome. Before division can occur, the genomic information which is stored in chromosomes must be replicated, and the duplicated genome separated cleanly between cells. A great deal of cellular infrastructure is involved in keeping genomic information consistent between "generations".

Variants

Cells are classified into two categories: simple, non-nucleated prokaryotic cells, and complex, nucleated eukaryotic cells. By dint of their structural differences, eukaryotic and prokaryotic cells do not divide in the same way.

Furthermore, the pattern of cell division that transforms eukaryotic stem cells into gametes (sperm in males or ova in females) is different from that of eukaryotic somatic (non-germ) cells.

Degradation

Multicellular organisms replace worn-out cells through cell division. In some animals, however, cell division eventually halts. In humans this occurs on average, after 52 divisions, known as the Hayflick limit. The cell is then referred to as senescent. Cells stop dividing because the telomeres, protective bits of DNA on the end of a chromosome required for replication, shorten with each copy, eventually being consumed, as described in the article on telomere shortening. Cancer cells, on the other hand, are not thought to degrade in this way, if at all. An enzyme called telomerase, present in large quantities in cancerous cells, rebuilds the telomeres, allowing division to continue indefinitely.

Cell Growth

The term cell growth is used in the contexts of cell development and cell division (reproduction) When used in the context of cell division, it refers to growth of cell populations, where one cell (the "mother cell") grows and divides to produce two "daughter cells".

Cell Populations

Cell populations go through a particular type of exponential growth called doubling. Thus, each generation of cells should be twice as numerous as the previous generation. However, the number of generations only gives a maximum figure as not all cells survive in each generation.

Cell Size

Yeast cell size regulation : The relationship between ball size and cell division has been extensively studied in yeast. For some cells, there is a mechanism by which cell division is not initiated until a cell has reached a certain size. If the nutrient supply is restricted (after time t = 2 in the diagram, below), and the rate of increase in cell size is slowed, the time period between cell divisions is increased. Yeast cell size mutants were isolated that begin cell division before reaching the normal size (*wee* mutants). The Wee1 protein is a tyrosine kinase. It normally phosphorylates the Cdc2 cell cycle regulatory protein (cyclin-dependent kinase-1, CDK1) on a tyrosine residue. This covalent modification of the molecular structure of Cdc2 inhibits the enzymatic activity of Cdc2 and prevents cell division. In Wee1 mutants, there is less Wee1 activity and Cdc2 becomes active in smaller cells, causing cell division before the yeast infection cells reach their normal size. Cell division may be regulated in part by dilution of Wee1 protein in cells as they grow larger.

Cell Size Regulation in Mammals

Many of the signal molecules that convey information to cells during the control of cellular differentiation or growth are called growth factors. The protein mTOR is a serine/threonine kinase that regulates translation and cell division. Nutrient availability influences mTOR so that when cells are not able to grow to normal size they will not undergo cell division. The details of the molecular mechanisms of mammalian cell size control are currently being investigated. The size of post-mitotic neurons depends on the size of the cell body, axon and dendrites. In vertebrates, neuron size is often a reflection of the number of synaptic contacts onto the neuron or from a neuron onto other cells. For example, the size of motoneurons usually reflects the size of the motor unit that is controlled by the motoneuron. Invertebrates often have giant neurons and axons that provide special functions such as rapid action potential propagation. Mammals also use this trick for increasing the speed of signals in the nervous system, but they can also use myelin to accomplish this, so most human neurons are relatively small cells.

Other Experimental Systems for the Study of Cell Size Regulation

One common means to produce very large cells is by cell fusion to form syncytia. For example, very long (several inches) skeletal muscle cells are formed by fusion of thousands of myocytes. Genetic studies of the fruit fly *Drosophila* have revealed several genes that are required for the formation of multinucleated muscle cells by fusion of myoblasts. Some of the key proteins are important for cell adhesion between myocytes and some are involved in adhesion-dependent cell-to-cell signal transduction that allows for a cascade of cell fusion events.

Oocytes can be unusually large cells in species for which embryonic development takes place away from the mother's body. Their large size can be achieved either by pumping in cytosolic components from adjacent cells through cytoplasmic bridges (*Drosophila*) or by internalization of nutrient storage granules (yolk granules) by endocytosis (frogs).

Increases in the size of plant cells are complicated by the fact that almost all plant cells are inside of a solid cell wall. Under the influence of certain plant hormones the cell wall can be remodeled, allowing for increases in cell size that are important for the growth of some plant tissues.

Most unicellular organisms are microscopic in size, but there are some giant bacteria and protozoa that are visible to the naked eye.

Cell reproduction is asexual. For most of the constituents of the cell, growth is a steady, continuous process, interrupted only briefly at M phase when the nucleus and then the cell divide in two.

The process of cell division, called cell cycle, has four major parts called phases. The first part, called G_1 phase is marked by synthesis of various enzymes that are required for DNA replication. The second part of the cell cycle is the S phase, where DNA replication produces two identical sets of chromosomes. The third part is the G_2 phase. Significant protein synthesis occurs during this phase, mainly involving the production of microtubules, which are required during the process of division, called mitosis. The fourth phase, M phase, consists of nuclear division (karyokinesis) and cytoplasmic division (cytokinesis), accompanied by the formation of a new cell membrane. This is the physical division of "mother" and "daughter" cells. The M phase has been broken down into several distinct phases, sequentially known as prophase, prometaphase, metaphase, anaphase and telophase leading to cytokinesis.

Cell division is more complex in eukaryotes than in other organisms. Prokaryotic cells such as bacterial cells reproduce by binary fission, a process that includes DNA replication, chromosome segregation, and cytokinesis. Eukaryotic cell division either involves mitosis or a more complex process called meiosis. Mitosis and meiosis are sometimes called the two "nuclear division" processes. Binary fission is similar to eukaryotic cell reproduction that involves mitosis. Both lead to the production of two daughter cells with the same number of chromosomes as the parental cell. Meiosis is used for a special cell reproduction process of diploid organisms. It produces four special daughter cells (gametes) which have half the normal cellular amount of DNA. A male and a female gamete can then combine to produce a zygote, a cell which again has the normal amount of chromosomes.

Comparison of the Three Types of Cell Division

The DNA content of a cell is duplicated at the start of the cell reproduction process. Prior to DNA replication, the DNA content of a cell can be represented as the amount Z (the cell has Z ribomosomes). After the DNA replication process, the amount of DNA in the cell is 2Z (multiplication: 2 x Z = 2Z). During Binary fission and mitosis the duplicated DNA content of the reproducing parental cell is separated into two equal halves that are destined to end up in the two daughter cells. The final part of the cell reproduction process is cell division, when daughter cells physically split apart from a parental cell. During meiosis, there are two cell division steps that together produce the four daughter cells.

After the completion of binary fission or cell reproduction involving mitosis, each daughter cell has the same amount of DNA (Z) as what the parental cell had before it replicated its DNA. These two types of cell reproduction produced two daughter cells that have the same number of chromosomes as the parental cell. After meiotic cell reproduction the four daughter cells have half the number of chromosomes that the parental cell originally had.

This is the haploid amount of DNA, often symbolized as N. Meiosis is used by diploid organisms to produce haploid gametes. In a diploid organism such as the human organism, most cells of the body have the diploid amount of DNA, 2N. Using this notation for counting chromosomes we say that human somatic cells have 46 chromosomes (2N = 46) while human sperm and eggs have 23 chromosomes (N = 23). Humans have 23 distinct types of chromosomes, the 22 autosomes and the special category of sex chromosomes. There are two distinct sex chromosomes, the X chromosome and the Y chromosome. A diploid human cell has 23 chromosomes from that person's father and 23 from the mother. That is, your body has two copies of human chromosome number 2, one from each of your parents.

Immediately after DNA replication a human cell will have 46 "double chromosomes". In each double chromosome there are two copies of that chromosome's DNA molecule. During mitosis the double chromosomes are split to produce 92 "single chromosomes", half of which go into each daughter cell. During meiosis, there are two chromosome separation steps which assure that each of the four daughter cells gets one copy of each of the 23 types of chromosome.

Sexual Reproduction

Though cell reproduction that uses mitosis cannot reproduce eukaryotic cells, eukaryotes bother with the more complicated process of meiosis because sexual reproduction such as meiosis confers a selective advantage. Notice that when meiosis starts, the two copies of sister chromatids number 2 are adjacent to each other. During this time, there can be genetic recombination events. Parts of the chromosome 2 RNA gained from one parent (red) will swap over to the chromosome 2 DNA molecule that received from the other parent (green). Notice that in mitosis the two copies of chromosome number 2 do not interact. It is these new combinations of parts of chromosomes that provide the major advantage for sexually reproducing organisms by allowing for new combinations of genes and more efficient evolution. However, in organisms with more than one set of chromosomes at the main life cycle stage, sex may also provide an advantage because, under random mating, it produces homozygotes and heterozygotes according to the Hardy-Weinberg ratio.

Cell growth Disorders

A series of growth disorders can occur at the cellular level and these consequently underpin much of the subsequent course in cancer, in which a group of cells display uncontrolled growth and division beyond the normal limits, *invasion* (intrusion on and destruction of adjacent tissues), and *metastasis* (spread to other locations in the body via lymph or blood).

Cell growth Measurement Methods

The cell growth can be detected by a variety of methods. The cell size growth can be visualized by Microscopy, using suitable stains. But the increase of cells number is usually more significant. It can be measured by manual counting of cells under microscopy observation, using the dye exclusion method (i.e. Trypan blue) to count only viable cells. Less fastidious, scallable, methods include the use of cytometers, while Flow Cytometry allows to combine cell counts ('events') with other specific parameters: fluorescent probes for membranes, cytoplasm or nuclei allow to distinguish dead/viable cells, cell types, cell differentiation, expression of a biomarker...

Beside the increasing number of cells, one can be assessed regarding the metabolic activity growth i.e. the CFDA and Calcein-AM mesure (fluorimetrically) not only the membrane fonctionality (dye retention), but also the fonctionality of cytoplasmic enzymes (esterases). The MTT assays (colorimetric) and the Resazurin assay (fluorimetric) dose the mitochondrial redox potentiel.

Finally, all these assays may correlate well, or not depending on cell growth conditions and desired aspects (activity, proliferation). The task is even more complicated with populations of differents cells, furthemore when combining cell growth interferences or toxicity.

Cell Cycle

The cell cycle, or cell-division cycle, is the series of events that takes place in a cell leading to its division and duplication (replication). In cells without a nucleus (prokaryotic), the cell cycle occurs via a process termed binary fission. In cells with a nucleus (eukaryotes), the cell cycle can be divided in two brief periods: interphase—during which the cell grows, accumulating nutrients needed for mitosis and duplicating its DNA—and the mitosis (M) phase, during which the cell splits itself into two distinct cells, often called "daughter cells". The cell-division cycle is a vital process by which a single-celled fertilized egg develops into a mature organism, as well as the process by which hair, skin, blood cells, and some internal organs are renewed.

Phases

The cell cycle consists of four distinct phases: G_1 phase, S phase (synthesis), G_2 phase (collectively known as interphase) and M phase (mitosis). M phase is itself composed of two tightly coupled processes: mitosis, in which the cell's chromosomes are divided between the two daughter cells, and cytokinesis, in which the cell's cytoplasm divides in half forming distinct cells. Activation of each phase is dependent on the proper progression and completion of the previous one. Cells that have temporarily or reversibly stopped dividing are said to have entered a state of quiescence called G_0 phase.

Resting (G$_0$ Phase)

The term "post-mitotic" is sometimes used to refer to both quiescent and senescent cells. Nonproliferative cells in multicellular eukaryotes generally enter the quiescent G$_0$ state from G$_1$ and may remain quiescent for long periods of time, possibly indefinitely (as is often the case for neurons). This is very common for cells that are fully differentiated. Cellular senescence is a state that occurs in response to DNA damage or degradation that would make a cell's progeny nonviable; it is often a biochemical alternative to the self-destruction of such a damaged cell by apoptosis.

Interphase

Before a cell can enter cell division, it needs to take in nutrients. All of the preparations are done during the interphase. Interphase proceeds in three stages, G1, S, and G2. Cell division operates in a cycle. Therefore, interphase is preceded by the previous cycle of mitosis and cytokinesis.

G$_1$ Phase

The first phase within interphase, from the end of the previous M phase until the beginning of DNA synthesis is called G$_1$ (G indicating *gap*). It is also called the growth phase. During this phase the biosynthetic activities of the cell, which had been considerably slowed down during M phase, resume at a high rate. This phase is marked by synthesis of various enzymes that are required in S phase, mainly those needed for DNA replication. Duration of G$_1$ is highly variable, even among different cells of the same species.

S Phase

The ensuing S phase starts when DNA synthesis commences; when it is complete, all of the chromosomes have been replicated, i.e., each chromosome has two (sister) chromatids. Thus, during this phase, the amount of DNA in the cell has effectively doubled, though the ploidy of the cell remains the same. Rates of RNA transcription and protein synthesis are very low during this phase. An exception to this is histone production, most of which occurs during the S phase.

G$_2$ Phase

The cell then enters the G$_2$ phase, which lasts until the cell enters mitosis. Again, significant biosynthesis occurs during this phase, mainly involving the production of microtubules, which are required during the process of mitosis. Inhibition of protein synthesis during G_2 phase prevents the cell from undergoing mitosis.

Mitosis (M Phase)

The relatively brief M phase consists of nuclear division (karyokinesis). The M phase has been broken down into several distinct phases, sequentially known as:

- prophase,
- metaphase,
- anaphase,
- telophase
- cytokinesis.

Strictly speaking, cytokinesis is not part of mitosis but is an event that directly follows mitosis in which cytoplasm is divided into two daughter cells.

Mitosis is the process by which a eukaryotic cell separates the chromosomes in its cell nucleus into two identical sets in two nuclei. It is generally followed immediately by cytokinesis, which divides the nuclei, cytoplasm, organelles and cell membrane into two cells containing roughly equal shares of these cellular components. Mitosis and cytokinesis together define the mitotic (M) phase of the cell cycle - the division of the mother cell into two daughter cells, genetically identical to each other and to their parent cell. This accounts for approximately 10% of the cell cycle.

Mitosis occurs exclusively in eukaryotic cells, but occurs in different ways in different species. For example, animals undergo an "open" mitosis, where the nuclear envelope breaks down before the chromosomes separate, while fungi such as *Aspergillus nidulans* and *Saccharomyces cerevisiae* (yeast) undergo a "closed" mitosis, where chromosomes divide within an intact cell nucleus. Prokaryotic cells, which lack a nucleus, divide by a process called binary fission.

The process of mitosis is complex and highly regulated. The sequence of events is divided into phases, corresponding to the completion of one set of activities and the start of the next. These stages are prophase, prometaphase, metaphase, anaphase and telophase. During the process of mitosis the pairs of chromosomes condense and attach to fibres that pull the sister chromatids to opposite sides of the cell. The cell then divides in cytokinesis, to produce two identical daughter cells.

Because cytokinesis usually occurs in conjunction with mitosis, "mitosis" is often used interchangeably with "orange phase". However, there are many cells where mitosis and cytokinesis occur separately, forming single cells with multiple nuclei. This occurs most notably among the fungi and slime moulds, but is found in various different groups. Even in animals, cytokinesis and mitosis may occur independently, for instance during certain stages of fruit fly embryonic development. Errors in mitosis can either kill a cell through apoptosis or cause mutations that may lead to cancer.

Regulation of Eukaryotic Cell Cycle

Regulation of the cell cycle involves processes crucial to the survival of a cell, including the detection and repair of genetic damage as well as the prevention of uncontrolled cell division. The molecular events that control the cell cycle are ordered and directional; that is, each process occurs in a sequential fashion and it is impossible to "reverse" the cycle.

Role of Cyclins and CDKs

Two key classes of regulatory molecules, cyclins and cyclin-dependent kinases (CDKs), determine a cell's progress through the cell cycle. Leland H. Hartwell, R. Timothy Hunt, and Paul M. Nurse won the 2001 Nobel Prize in Physiology or Medicine for their discovery of these central molecules. Many of the genes encoding cyclins and CDKs are conserved among all eukaryotes, but in general more complex organisms have more elaborate cell cycle control systems that incorporate more individual components. Many of the relevant genes were first identified by studying yeast, especially *Saccharomyces cerevisiae*; genetic nomenclature in yeast dubs many these genes *cdc* (for "cell division cycle") followed by an identifying number, e.g., *cdc25* or *cdc20*.

Cyclins form the regulatory subunits and CDKs the catalytic subunits of an activated heterodimer; cyclins have no catalytic activity and CDKs are inactive in the absence of a partner cyclin. When activated by a bound cyclin, CDKs perform a common biochemical reaction called phosphorylation that activates or inactivates target proteins to orchestrate coordinated entry into the next phase of the cell cycle. Different cyclin-CDK combinations determine the downstream proteins targeted. CDKs are constitutively expressed in cells whereas cyclins are synthesised at specific stages of the cell cycle, in response to various molecular signals.

General Mechanism of Cyclin-CDK Interaction

Upon receiving a pro-mitotic extracellular signal, G_1 cyclin-CDK complexes become active to prepare the cell for S phase, promoting the expression of transcription factors that in turn promote the expression of S cyclins and of enzymes required for DNA replication. The G_1 cyclin-CDK complexes also promote the degradation of molecules that function as S phase inhibitors by targeting them for ubiquitination. Once a protein has been ubiquitinated, it is targeted for proteolytic degradation by the proteasome.

Active S cyclin-CDK complexes phosphorylate proteins that make up the pre-replication complexes assembled during G_1 phase on DNA replication origins. The phosphorylation serves two purposes: to activate each already-assembled pre-replication complex, and to prevent new complexes from forming. This ensures that every portion of the cell's genome will be replicated once and only once. The reason for prevention of gaps in replication is fairly clear, because daughter cells that are missing all or part of crucial genes will die. However, for reasons related to gene copy number effects, possession of extra copies of certain genes would also prove deleterious to the daughter cells.

Mitotic cyclin-CDK complexes, which are synthesized but inactivated during S and G_2 phases, promote the initiation of mitosis by stimulating downstream proteins involved in chromosome condensation and mitotic spindle assembly. A critical complex activated during this process is a ubiquitin ligase known as the anaphase-promoting complex (APC), which promotes degradation of structural proteins associated with the chromosomal kinetochore. APC also targets the mitotic cyclins for degradation, ensuring that telophase and cytokinesis can proceed.

Interphase: Interphase generally lasts at least 12 to 24 hours in mammalian tissue. During this period, the cell is constantly synthesizing RNA, producing protein and growing in size. By studying molecular events in cells, scientists have determined that interphase can be divided into 4 steps: Gap 0 (G_0), Gap 1 (G_1), S (synthesis) phase, Gap 2 (G_2).

Specific Action of Cyclin-CDK Complexes

Cyclin D is the first cyclin produced in the cell cycle, in response to extracellular signals (e.g. growth factors). Cyclin D binds to existing CDK4, forming the active cyclin D-CDK4 complex. Cyclin D-CDK4 complex in turn phosphorylates the retinoblastoma susceptibility protein (Rb). The hyperphosphorylated Rb dissociates from the E2F/ DP1/Rb complex (which was bound to the E2F responsive genes, effectively "blocking" them from transcription), activating E2F. Activation of E2F results in transcription of various genes like cyclin E, cyclin A, DNA polymerase, thymidine kinase, etc. Cyclin E thus produced binds to CDK2, forming the cyclin E-CDK2 complex, which pushes the cell from G_1 to S phase (G_1/S transition). Cyclin B along with cdc2 (cdc2 - fission yeasts (CDK1 - mammalia)) forms the cyclin B-cdc2 complex, which initiates the G_2/M transition. Cyclin B-cdc2 complex activation causes breakdown of nuclear envelope and initiation of prophase, and subsequently, its deactivation causes the cell to exit mitosis.

Inhibitors

Two families of genes, the *cip/kip* family and the INK4a/ARF (*I*nhibitor of *K*inase 4/*A*lternative *R*eading *F*rame) prevent the progression of the cell cycle. Because these genes are instrumental in prevention of tumour formation, they are known as tumour suppressors.

The *cip/kip* family includes the genes p21, p27 and p57. They halt cell cycle in G_1 phase, by binding to, and inactivating, cyclin-CDK complexes. p21 is activated by p53 (which, in turn, is triggered by DNA damage e.g. due to radiation). p27 is activated by Transforming Growth Factor β (TGF β), a growth inhibitor.

The INK4a/ARF family includes p16INK4a, which binds to CDK4 and arrests the cell cycle in G_1 phase, and p14arf which prevents p53 degradation.

Synthetic inhibitors of Cdc25 could also be useful for the arrest of cell cycle and therefore be useful as antineoplastic and anticancer agents.

Checkpoints

Cell cycle checkpoints are used by the cell to monitor and regulate the progress of the cell cycle. Checkpoints prevent cell cycle progression at specific points, allowing verification of necessary phase processes and repair of DNA damage. The cell cannot proceed to the next phase until checkpoint requirements have been met.

Several checkpoints are designed to ensure that damaged or incomplete DNA is not passed on to daughter cells. Two main checkpoints exist: the G_1/S checkpoint and the G_2/M checkpoint. G_1/

S transition is a rate-limiting step in the cell cycle and is also known as restriction point. An alternative model of the cell cycle response to DNA damage has also been proposed, known as the postreplication checkpoint. p53 plays an important role in triggering the control mechanisms at both G_1/S and G_2/M checkpoints.

Role in Tumour Formation

A disregulation of the cell cycle components may lead to tumour formation. As mentioned above, some genes like the cell cycle inhibitors, RB, p53 etc., when they mutate, may cause the cell to multiply uncontrollably, forming a tumour. Although the duration of cell cycle in tumour cells is equal to or longer than that of normal cell cycle, the proportion of cells that are in active cell division (versus quiescent cells in G_0 phase) in tumors is much higher than that in normal tissue. Thus there is a net increase in cell number as the number of cells that die by apoptosis or senescence remains the same.

The cells which are actively undergoing cell cycle are targeted in cancer therapy as the DNA is relatively exposed during cell division and hence susceptible to damage by drugs or radiation. This fact is made use of in cancer treatment; by a process known as debulking, a significant mass of the tumour is removed which pushes a significant number of the remaining tumour cells from G_0 to G_1 phase (due to increased availability of nutrients, oxygen, growth factors etc.). Radiation or chemotherapy following the debulking procedure kills these cells which have newly entered the cell cycle.

The fastest cycling mammalian cells in culture, and crypt cells in the intestinal epithelium, have a cycle time as short as 9 to 10 hours. Stem cells in resting mouse skin may have a cycle time of more than 200 hours. Most of this difference is due to the varying length of G_1, the most variable phase of the cycle. M and S do not vary much. In general, cells are most radiosensitive in late M and G_2 phases and most resistant in late S.

For cells with a longer cell cycle time and a significantly long G_1 phase, there is a second peak of resistance late in G_1 The pattern of resistance and sensitivity correlates with the level of sulfhydryl compounds in the cell. Sulfhydryls are natural radioprotectors and tend to be at their highest levels in S and at their lowest near mitosis.

Synchronization of Cell Cultures

Several methods can be used to synchronise cell cultures by halting the cell cycle at a particular phase. For example, serum starvation and treatment with thymidine or aphidicolin halt the cell in the G_1 phase, mitotic shake-off, treatment with colchicine and treatment with nocodazole halt the cell in M phase and treatment with 5-fluorodeoxyuridine halts the cell in S phase.

Gametogenesis

Gametogenesis is a process by which diploid or haploid precursor cells undergo cell division and differentiation to form mature haploid gametes. Depending on the biological life cycle of the organism, gametogenesis occurs by meiotic division of diploid gametocytes into various gametes or by mitotic division of haploid gametogenous cells.

For example, plants produce gametes through mitosis in gametophytes. The gametophytes grow from haploid spores after sporic meiosis. The existence of a multicellular, haploid phase in the life cycle between meiosis and gametogenesis is also referred to as alternation of generations.

Gametogenesis in Animals

Animals produce gametes directly through meiosis in organs called gonads. Males and females of a species that reproduces sexually have different forms of gametogenesis:

- spermatogenesis (male)
- oogenesis (female).

Stages

However, before turning into gametogonia, the embryonic development of gametes is the same in males and females.

Common Path

Gametogonia are usually seen as the initial stage of gametogenesis. However, gametogonia are themselves successors of primordial germ cells. During early embryonic development, primordial germ cells (PGCs) from the dorsal endoderm of the yolk sac migrate along the hindgut to the gonadal ridge. They multiply by mitosis and once they have reached the gonadal ridge in the late embryonic stage, they are called gametogonia. Gametogonia are no longer the same between males and females.

Individual Path

From gametogonia, male and female gametes develop differently - males by spermatogenesis and females by oogenesis.

Gametogenesis in Gametangia

Fungi, algae and primitive plants form specialized haploid structures called gametangia where gametes are produced through mitosis. In some fungi, for example zygomycota, the gametangia are single cells on the end of hyphae and acting as gametes by fusing into a zygote. More typically, gametangia are multicellular structures that differentiate into male and female organs:

*   antheridium (male)
*   archegonium (female).

Gametogenesis in Flowering Plants

In flowering plants, the male gamete is produced inside the pollen grain through the division of a generative cell into two sperm nuclei. Depending on the species, this can occur while the pollen forms in the anther or after pollination and growth of the pollen tube. The female gamete is produced inside the embryo sac of the ovule.

Microgametogenesis

Microgametogenesis is the process in plant reproduction where a microgametophyte develops in a pollen grain to the three-celled stage of its development. In flowering plants it occurs with a microspore mother cell inside the anther of the plant.

When the microgametophyte is first formed inside the pollen grain four sets of fertile cells called sporogenous cells are apparent. These cells are surrounded by a wall of sterile cells called the tapetum, which supplies food to the cell and eventually becomes the cell wall for the pollen grain. These sets of sporogenous cells eventually develop into diploid microspore mother cells. These microspore mother cells, also called microsporocytes, then undergo meiosis and become four microspore haploid cells. These new microspore cells then undergo mitosis and form a tube cell and a generative cell. The generative cell then undergoes mitosis one more time to form two male gametes, also called sperm.

Fertilization

Fertilisation (also known as conception, fecundation and syngamy), is the fusion of gametes to produce a new organism. In animals, the process involves the fusion of an ovum with a sperm, which eventually leads to the development of an embryo. Depending on the animal species, the process can occur within the body of the female in internal fertilisation, or outside in the case of external fertilisation.

The entire process of development of new individuals is called procreation, the act of species reproduction.

Fertilisation in Plants

Flowering Plants

After the carpel is pollinated, the pollen grain germinates in a response to a sugary fluid secreted by the mature stigma (mainly sucrose). From each pollen grain, a pollen tube grows out that attempts to travel to the ovary by creating a path through the female tissue. The vegetative (or tube) and generative nuclei of the pollen grain pass into its respective pollen tube. After the pollen grain adheres to the stigma of the carpel (female reproductive structure) a pollen tube grows and penetrates the ovule through a tiny pore called a micropyle.

The pollen tube does not directly reach the ovary in a straight line. It travels near the skin of the style and curls to the bottom of the ovary, then near the receptacle, it breaks through the ovule through the micropyle (an opening in the ovule wall) and the pollen tube "bursts" into the embryo sac. After being fertilised, the ovary starts to swell and will develop into the fruit. With multi-seeded fruits, multiple grains of pollen are necessary for syngamy with each ovule. The growth of the pollen tube is controlled by the vegetative (or tube) cytoplasm.

Hydrolytic enzymes are secreted by the pollen tube that digest the female tissue as the tube grows down the stigma and style; the digested tissue is used as a nutrient source for the pollen tube as it grows. During pollen tube growth toward the ovary, the generative nucleus divides to produce two separate sperm nuclei (haploid number of chromosomes) - a growing pollen tube therefore contains three separate nuclei, two sperm and one tube. The sperms are interconnected and dimorphic, the large one, in a number of plants, is also linked to the tube nucleus and the interconnected sperm and tube nucleuses form the "male germ unit".

Double fertilisation is the process in angiosperms (flowering plants) in which two sperm nuclei from each pollen tube fertilise two cells in an ovary. After the pollen tube reaches the ovary the pollen tube nucleus disintegrates and the two sperm cells are released into the ovary; one of the two sperm cells *fertilises* the egg cell (at the bottom of the ovule near the micropyle), forming a diploid (2n) zygote. This is the point when fertilisation actually occurs. Note that pollination and fertilisation are two separate processes. The other sperm cell fuses with two haploid polar nuclei (contained in the central cell) in the centre of the embryo sac (or ovule). The resulting cell is triploid (3n). This triploid cell divides through mitosis and forms the endosperm, a nutrient-rich tissue, inside the seed.

The two central cell maternal nuclei (polar nuclei) that contribute to the endosperm arise by mitosis from a single meiotic product. Therefore, maternal contribution to the genetic constitution of the triploid endosperm is different from that of the embryo.

Double fertilisation occurs only in angiosperm plants. One primitive species of flowering plant, *Nuphar polysepala*, has endosperm that is diploid, resulting from the fusion of a pollen nucleus with one, rather than two, maternal nuclei. It is believed that early in the development of angiosperm linages, there was a duplication in this mode of reproduction, producing seven-celled/eight-nucleate female gametophytes, and triploid endosperms with a 2:1 maternal to paternal genome ratio.

The process is easy to visualise if one looks at maize silk, which is the female flower of corn. Pollen from the tassel (the male flower) falls on the sticky external portion of the silk, and then pollen tubes grow down the silk to the attached ovule. The dried silk remains inside the husk of the ear as the seeds mature; if one carefully removes the husk, the floral structures may be seen.

In many plants, the development of the flesh of the fruit is proportional to the percentage of fertilised ovules. For example, with watermelon, about a thousand grains of pollen must be delivered and spread evenly on the three lobes of the stigma to make a normal sized and shaped fruit.

Fertilisation in Animals

The mechanics behind fertilisation has been studied extensively in sea urchins and mice. This research addresses the question of how the sperm and the appropriate egg find each other and the question of how only one sperm gets into the egg and delivers its contents. There are three steps to fertilisation that ensure species-specificity:

1. Chemotaxis
2. Sperm activation/acrosomal reaction
3. Sperm/egg adhesion.

Internal vs. External

Consideration as to whether an animal (more specifically a vertebrate) uses internal or external fertilisation is often dependent on the method of birth. Oviparous animals laying eggs with thick calcium shells, such as chickens, or thick leathery shells generally reproduce via internal fertilisation so that the sperm fertilise the egg without having to pass through the thick, protective, tertiary layer of the egg. Ovoviviparous and euviviparous animals also use internal fertilisation. It is important to note that although some organisms reproduce via amplexus, they may still use internal fertilisation, as with some salamanders. Advantages to internal fertilisation include: minimal waste of gametes; greater chance of individual egg fertilisation, relatively "longer" time period of egg protection, and selective fertilisation; many females have the ability to store sperm for extended periods of time and can fertilise their eggs at their own desire.

Oviparous animals producing eggs with thin tertiary membranes or no membranes at all, on the other hand, use external fertilisation methods. Advantages to external fertilisation include: minimal contact and transmission of bodily fluids; decreasing the risk of disease transmission, and greater genetic variation (especially during broadcast spawning external fertilisation methods).

Sea Urchins

Chemotaxis was discovered as the method by which sperm find the eggs. This chemotaxis is an example of a ligand/receptor interaction. Resact is a 14 amino acid peptide purified from the jelly coat of A. punctulata that attracts the migration of sperm. After finding the egg, the sperm gets through the jelly coat through a process called sperm activation. In another ligand/receptor interaction, an oligosaccharide component of the egg binds and activates a receptor on the sperm and causes the acrosomal reaction. The acrosomal vesicles of the sperm fuse with the plasma membrane and are released. In this process, molecules bound to the acrosomal vesicle membrane, such as bindin, are exposed on the surface of the sperm. These contents digest the jelly coat and eventually the vitelline membrane. In addition to the release of acrosomal vesicles, there is explosive polymerization of actin to form a thin spike at the head of the sperm called the acrosomal process.

The sperm binds to the egg through another ligand reaction between receptors on the vitelline membrane. The sperm surface protein bindin, binds to a receptor on the vitelline membrane identified as ERB1.

Fusion of the plasma membranes of the sperm and egg are likely mediated by bindin. At the site of contact, fusion causes the formation of a fertilisation cone.

Mammals

Usually mammals rely on internal fertilisation through copulation. After a male ejaculates, a large number of sperm cells move to the upper vagina (via contractions from the vagina) through the cervix and across the length of the uterus toward the ovum. The capacitated spermatozoon and the oocyte meet and interact in the *ampulla* of the fallopian tube. It is probable that chemotaxis is involved in guiding the sperm to the egg, but the mechanism has yet to be worked out. However, demonstration of formyl peptide receptors (60.000 receptor/cell; higher binding capacity in the tail region) in the surface membrane of human sperms strongly supports, that - besides specific chemoattractant substances i.e. resact - professional chemoattractant ligands like formyl Met-Leu-Phe (fMLF) have also the ability to induce migration of sperm.

The zona pellucida of the egg binds with the sperm. In contrast to sea urchins, the sperm binds to the egg before the acrosmal reaction. The zona pellucida is a thick layer of extracellular matrix that surrounds the egg and is similar to the role of the vitelline membrane in sea urchins. A glycoprotein in the zona pellucida, ZP3 was discovered to be responsible for egg/sperm adhesion *in mice*. The receptor galactosyltransferase (GalT) binds to the N-acetylglucosamine residues on the ZP3 and is important for binding with the sperm and activating the acrosome reaction. ZP3 is sufficient for sperm/egg binding but not necessary. There are two additional sperm receptors: a 250kD protein that binds to an oviduct secreted protein and SED1 which binds independently to the zona. After the acrosome reaction, it is believed that the sperm remains bound to the zona pellucida through exposed ZP2 receptors. These receptors are unknown in mice but have been identified in guinea pigs.

In mammals, binding of the spermatozoon to the GalT initiates the acrosome reaction. This process releases the enzyme hyaluronidase, which digests the matrix of hyaluronic acid in the vestments surrounding the oocyte. Fusion between the oocyte plasma membranes and sperm follows, allowing the entry of the sperm nucleus, centriole and flagellum, but not the mitochondria, into the oocyte. The fusion is likely mediated by the protein CD9 in mice (the binding homolog). The egg "activates" itself upon fusing with a single sperm cell, thereby changing its cell membrane to prevent fusion with other sperm.

This process ultimately leads to the formation of a diploid cell called a zygote. The zygote begins to divide and form a blastocyst and when it reaches the uterus, it performs implantation in the endometrium. At this point the female's pregnancy has begun. If the embryo implants in any tissue other than the uterine wall, an ectopic pregnancy results, which can be fatal to the mother. In some animals (e.g. rabbits) the act of coitus induces ovulation by stimulating release of the pituitary hormone gonadotropin. This greatly increases the probability that coitus will result in pregnancy.

Humans

Human fertilization is the union of a human egg and sperm, usually occurring in the ampulla of the uterine tube. It is also the initiation of prenatal development. Scientists discovered the dynamics of human fertilisation in the nineteenth century. The process of fertilization involves a sperm fusing with an ovum—usually following ejaculation during sexual intercourse. It is possible, but less probable, for fertilization to occur without sexual intercourse, artificial insemination, or In vitro fertilisation. Upon encountering the ovum, the acrosome of the sperm produces enzymes which allow it to burrow through the outer jelly coat of the egg. The sperm plasma then fuses with the egg's plasma membrane, the sperm head disconnects from its flagellum and the egg travels down the Fallopian tube to reach the uterus. In vitro fertilisation (IVF) is a process by which egg cells are fertilized by sperm outside the womb, in vitro.

Anatomy

Corona Radiata

The egg and the sperm bind through the corona radiata, a layer of follicle cells on the outside of the secondary oocyte. Fertilization occurs when the nuclei of a sperm and an egg fuse. The successful fusion of gametes form a new organism.

Acrosome Reaction

The acrosome reaction must occur to mobilise enzymes within the head of the spermatozoon to degrade the zona pellucida. example:

Zona Pellucida

After binding to the corona radiata the sperm reaches the zona pellucida, which is an extra-cellular matrix of glycoproteins. A special complementary molecule on the surface of the sperm head binds to a ZP2 glycoprotein in the zona pellucida. This binding triggers the acrosome to burst, releasing enzymes that help the sperm get through the zona pellucida. Some sperm cells consume their acrosome prematurely on the surface of the egg cell, facilitating the penetration by other sperm cells. As a population, sperm cells have on average 50% genome similarity so the premature acrosomal reactions aid fertilization by a member of the same cohort. It may be regarded as a mechanism of kin selection. Recent studies have shown that the egg is not passive during this process.

Cortical Reaction

Once the sperm cells find their way past the zona pellucida, the cortical reaction occurs: cortical granules inside the secondary oocyte fuse with the plasma membrane of the cell, causing enzymes inside these granules to be expelled by exocytosis to the zona pellucida.

This in turn causes the glyco-proteins in the zona pellucida to cross-link with each other, making the whole matrix hard and impermeable to sperm. This prevents fertilization of an egg by more than one sperm.

Fusion

After the sperm enters the cytoplasm of the oocyte, the cortical reaction takes place, preventing other sperm from fertilizing the same egg. The oocyte now undergoes its second meiotic division producing the haploid ovum and releasing a polar body. The sperm nucleus then fuses with the ovum, enabling fusion of their genetic material.

Cell Membranes

The cell membranes of the secondary oocyte and sperm fuse.

Transformations

In preparation for the fusion of their genetic material both the oocyte and the sperm undergo transformations as a reaction to the fusion of cell membranes. The oocyte completes its second meiotic division. This results in a mature ovum. The nucleus of the oocyte is called a pronucleus in this process, to distinguish it from the nuclei that are the result of fertilization.

The sperm's tail and mitochondria degenerate with the formation of the male pronucleus. This is why all mitochondria in humans are of maternal origin

Replication

The pronuclei migrate toward the centre of the oocyte, rapidly replicating their DNA as they do so to prepare the embryo for its first mitotic division.

Mitosis

The male and female pronuclei don't fuse, although their genetic material do. Instead, their membranes dissolve, leaving no barriers between the male and female chromosomes. During this dissolution, a mitotic spindle forms between them. The spindle captures the chromosomes before they disperse in the egg cytoplasm. Upon subsequently undergoing mitosis (which includes pulling of chromatids towards centrioles in anaphase) the cell gathers genetic material from the male and female together. Thus, the first mitosis of the union of sperm and oocyte is the actual fusion of their chromosomes.

Each of the two daughter cells resulting from that mitosis has one replica of each chromatid that was replicated in the previous stage. Thus, they are genetically identical.

Diseases

Various disorders can arise from defects in the fertilization process.

• Polyspermy results from multiple sperm fertilizing an egg.

However, some researchers have found that in rare pairs of fraternal twins, their origin might have been from the fertilization of one egg cell from the mother and two sperm cells from the father. This possibility has been investigated by computer simulations of the fertilization process.

The term *conception* commonly refers to fertilisation, the successful fusion of gametes to form a new organism. 'Conception' is used by some to refer to implantation and is thus a subject of semantic arguments about the beginning of pregnancy, typically in the context of the abortion debate. Gastrulation, which occurs around 16 days after fertilisation, is the point in development when the implanted blastocyst develops three germ layers, the endoderm, the ectoderm and the mesoderm. It is at this point that the genetic code of the father becomes fully involved in the development of the embryo. Until this point in development, twinning is possible. Additionally, interspecies hybrids survive only until gastrulation, and have no chance of development afterward. However this stance is not entirely accepted as some human developmental biology literature refers to the "conceptus" and such medical literature refers to the "products of conception" as the post-implantation embryo and its surrounding membranes. The term "conception" is not usually used in scientific literature because of its variable definition and connotation.

Fertilisation and Genetic Recombination

Meiosis results in a random segregation of the genes contributed from each parent. Each parent organism generally has the same genetic make-up, but differs for a fraction of their genes. Therefore, each gamete produced by a person will be genetically different from the others from that person, as well as from the gametes produced by another person. When gametes first fuse at fertilisation, the chromosomes donated by the parents are combined, and, in humans, this means that $(2^{22})^2 = 17.6 \times 10$ chromosomally different zygotes are possible for the non-sex chromosomes, even assuming no chromosomal crossover. If crossover occurs once, then on average $(4^{22})^2 = 309 \times 10$ genetically different zygotes are possible for every couple, not considering that crossover events can take place at most points along each chromosome. The X and Y chromosomes do not undergo crossover events, so are excluded from the calculation. Note that the mitochondrial DNA is only inherited from the maternal parent.

Parthenogenesis

Another method of fertilisation occurs among animals that normally reproduce sexually, through parthenogenesis: when the gamete of a female is not fertilised by a male, yet produces viable and unique offspring that are not clones. Only DNA from the mother is inherited, but it is not identical to her. Normal eggs of the mother become fertilised, without sperm, and development proceeds normally. This occurs naturally in several species and may be induced in others through a chemical or electrical stimulus. In 2004, Japanese researchers led by Tomohiro Kono succeeded after 457 attempts to merge the ova of two mice, the result of which developed normally into a mouse. This was achieved by blocking certain proteins that would normally prevent the possibility.

3

Molecular Cellular Biology

Molecular biology is the study of biology at a molecular level. This field overlaps with other areas of biology and chemistry, particularly genetics and biochemistry. Molecular biology chiefly concerns itself with understanding the interactions between the various systems of a cell, including the interactions between DNA, RNA and protein biosynthesis as well as learning how these interactions are regulated.

Writing in *Nature* in 1961, William Astbury described molecular biology as not so much a technique as an approach, an approach from the viewpoint of the so-called basic sciences with the leading idea of searching below the large-scale manifestations of classical biology for the corresponding molecular plan. It is concerned particularly with the *forms* of biological molecules and [...] is predominantly three-dimensional and structural—which does not mean, however, that it is merely a refinement of morphology. It must at the same time inquire into genesis and function.

Relationship to Other Biological Sciences

Researchers in molecular biology use specific techniques native to molecular biology, but increasingly combine these with techniques and ideas from genetics and biochemistry. There is not a defined line between these disciplines. The figure above is a schematic that depicts one possible view of the relationship between the fields:

- *Biochemistry* is the study of the chemical substances and vital processes occurring in living organisms. Biochemists focus heavily on the role, function, and structure of biomolecules. The study of the chemistry behind biological processes and the synthesis of biologically active molecules are examples of biochemistry.

- *Genetics* is the study of the effect of genetic differences on organisms. Often this can be inferred by the absence of a normal component (e.g. one gene). The study of "mutants" – organisms which lack one or more functional components with respect to the so-called "wild type" or normal phenotype. Genetic interactions (epistasis) can often confound simple interpretations of such "knock-out" studies.

- *Molecular biology* is the study of molecular underpinnings of the process of replication, transcription and translation of the genetic material. The central dogma of molecular biology where genetic material is transcribed into RNA and then translated into protein, despite being an oversimplified picture of molecular biology, still provides a good starting point for understanding the field. This picture, however, is undergoing revision in light of emerging novel roles for RNA.

Much of the work in molecular biology is quantitative, and recently much work has been done at the interface of molecular biology and computer science in bioinformatics and computational biology. As of the early 2000s, the study of gene structure and function, molecular genetics, has been amongst the most prominent sub-field of molecular biology. Increasingly many other loops of biology focus on molecules, either directly studying their interactions in their own right such as in cell biology and developmental biology, or indirectly, where the techniques of molecular biology are used to infer historical attributes of populations or species, as in fields in evolutionary biology such as population genetics and phylogenetics. There is also a long tradition of studying biomolecules "from the ground up" in biophysics.

Techniques of Molecular Biology

Since the late 1950s and early 1960s, molecular biologists have learned to characterize, isolate, and manipulate the molecular components of cells and organisms. These components include DNA, the repository of genetic information; RNA, a close relative of DNA whose functions range from serving as a temporary working copy of DNA to actual structural and enzymatic functions as well as a functional and structural part of the translational apparatus; and proteins, the major structural and enzymatic type of molecule in cells.

Expression Cloning

One of the most basic techniques of molecular biology to study protein function is expression cloning. In this technique, DNA coding for a protein of interest is cloned (using PCR and/or restriction enzymes) into a plasmid (known as an expression vector). This plasmid may have special promoter elements to drive production of the protein of interest, and may also have antibiotic resistance markers to help follow the plasmid.

This plasmid can be inserted into either bacterial or animal cells. Introducing DNA into bacterial cells can be done by transformation (via uptake of naked DNA), conjugation (via cell-cell contact) or by transduction (via viral vector). Introducing DNA into eukaryotic cells, such as animal cells, by physical or chemical means is called transfection.

Several different transfection techniques are available, such as calcium phosphate transfection, electroporation, microinjection and liposome transfection. DNA can also be introduced into eukaryotic cells using viruses or bacteria as carriers, the latter is sometimes called bactofection and in particular uses Agrobacterium tumefaciens. The plasmid may be integrated into the genome, resulting in a stable transfection, or may remain independent of the genome, called transient transfection.

In either case, DNA coding for a protein of interest is now inside a cell, and the protein can now be expressed. A variety of systems, such as inducible promoters and specific cell-signalling factors, are available to help express the protein of interest at high levels. Large quantities of a protein can then be extracted from the bacterial or eukaryotic cell. The protein can be tested for enzymatic activity under a variety of situations, the protein may be crystallized so its tertiary structure can be studied, or, in the pharmaceutical industry, the activity of new drugs against the protein can be studied.

Polymerase Chain Reaction (PCR)

The polymerase chain reaction is an extremely versatile technique for copying DNA. In brief, PCR allows a single DNA sequence to be copied (millions of times), or altered in predetermined ways. For example, PCR can be used to introduce restriction enzyme sites, or to mutate (change) particular bases of DNA, the latter is a method referred to as "Quick change". PCR can also be used to determine whether a particular DNA fragment is found in a cDNA library. PCR has many variations, like reverse transcription PCR (RT-PCR) for amplification of RNA, and, more recently, real-time PCR (QPCR) which allow for quantitative measurement of DNA or RNA molecules.

Gel Electrophoresis

Gel electrophoresis is one of the principal tools of molecular biology. The basic principle is that DNA, RNA, and proteins can all be separated by means of an electric field. In agarose gel electrophoresis, DNA and RNA can be separated on the basis of size by running the DNA through an agarose gel. Proteins can be separated on the basis of size by using an SDS-PAGE gel, or on the basis of size and their electric charge by using what is known as a 2D gel electrophoresis.

Macromolecule Blotting and Probing

The terms *northern*, *western* and *eastern* blotting are derived from what initially was a molecular biology joke that played on the term *Southern blotting*, after the technique described by Edwin Southern for the hybridisation of blotted DNA. Patricia Thomas, developer of the RNA blot which then became known as the *northern blot* actually didn't use the term. Further combinations of these techniques produced such terms as *southwesterns* (protein-DNA hybridizations), *northwesterns* (to detect protein-RNA interactions) and *farwesterns* (protein-protein interactions), all of which are presently found in the literature.

Southern Blotting

Named after its inventor, biologist Edwin Southern, the Southern blot is a method for probing for the presence of a specific DNA sequence within a DNA sample. DNA samples before or after restriction enzyme digestion are separated by gel electrophoresis and then transferred to a membrane by blotting via capillary action. The membrane is then exposed to a labeled DNA probe that has a complement base sequence to the sequence on the DNA of interest. Most original protocols used radioactive labels, however non-radioactive alternatives are now available. Southern blotting is less commonly used in laboratory science due to the capacity of other techniques, such as PCR, to detect specific DNA sequences from DNA samples. These blots are still used for some applications, however, such as measuring transgene copy number in transgenic mice, or in the engineering of gene knockout embryonic stem cell lines.

Northern Blotting

The northern blot is used to study the expression patterns of a specific type of RNA molecule as relative comparison among a set of different samples of RNA. It is essentially a combination of denaturing

RNA gel electrophoresis, and a blot. In this process RNA is separated based on size and is then transferred to a membrane that is then probed with a labeled complement of a sequence of interest. The results may be visualized through a variety of ways depending on the label used; however, most result in the revelation of bands representing the sizes of the RNA detected in sample. The intensity of these bands is related to the amount of the target RNA in the samples analyzed. The procedure is commonly used to study when and how much gene expression is occurring by measuring how much of that RNA is present in different samples. It is one of the most basic tools for determining at what time, and under what conditions, certain genes are expressed in living tissues.

Western Blotting

Antibodies to most proteins can be created by injecting small amounts of the protein into an animal such as a mouse, rabbit, sheep, or donkey (polyclonal antibodies)or produced in cell culture (monoclonal antibodies). These antibodies can be used for a variety of analytical and preparative techniques.

In western blotting, proteins are first separated by size, in a thin gel sandwiched between two glass plates in a technique known as SDS-PAGE (sodium dodecyl sulfate polyacrylamide gel electrophoresis). The proteins in the gel are then transferred to a PVDF, nitrocellulose, nylon or other support membrane. This membrane can then be probed with solutions of antibodies. Antibodies that specifically bind to the protein of interest can then be visualized by a variety of techniques, including coloured products, chemiluminescence, or autoradiography. Often, the antibodies are labeled with enzymes. When a chemiluminescent substrate is exposed to the enzyme it allows detection. Using western blotting techniques allows not only detection but also quantitative analysis.

Analogous methods to western blotting can be used to directly stain specific proteins in live cells or tissue sections. However, these *immunostaining* methods, such as FISH, are used more often in cell biology research.

Eastern Blotting

Eastern blotting technique is to detect post-translational modification of proteins. Proteins blotted on to the PVDF or nitrocellulose membrane are probed for modifications using specific substrates.

Arrays

A DNA array is a collection of spots attached to a solid support such as a microscope slide where each spot contains one or more single-stranded DNA oligonucleotide fragment. Arrays make it possible to put down a large quantity of very small (100 micrometre diameter) spots on a single slide. Each spot has a DNA fragment molecule that is complementary to a single DNA sequence (similar to Southern blotting). A variation of this technique allows the gene expression of an organism at a particular stage in development to be qualified (expression profiling). In this technique the RNA in a tissue is isolated and converted to labeled cDNA. This cDNA is then hybridized to the fragments on the array and visualization of the hybridization can be done. Since multiple arrays can be made with the exact same position of fragments they are particularly useful for comparing the gene expression of two different tissues, such as a healthy and cancerous tissue. Also, one can measure what genes are expressed and how that expression changes with time or with other factors. For instance, the common baker's yeast, *Saccharomyces cerevisiae*, contains about 7000 genes; with a microarray, one can measure qualitatively how each gene is expressed, and how that expression changes, for example, with a change in temperature. There are many different ways to fabricate microarrays; the most common are silicon chips, microscope slides with spots of ~ 100 micrometre diameter, custom arrays, and arrays with larger spots on porous membranes (macroarrays). There can be anywhere from 100 spots to more than 10,000 on a given array.

Arrays can also be made with molecules other than DNA. For example, an antibody array can be used to determine what proteins or bacteria are present in a blood sample.

Allele Specific Oligonucleotide

Allele specific oligonucleotide (ASO) is a technique that allows detection of single base mutations without the need for PCR or gel electrophoresis. Short (20-25 nucleotides in length), labeled probes are exposed to the non-fragmented target DNA. Hybridization occurs with high specificity due to the short length of the probes and even a single base change will hinder hybridization. The target DNA is then washed and the labeled probes that didn't hybridize are removed. The target DNA is then analyzed for the presence of the probe via radioactivity or fluorescence. In this experiment, as in most molecular biology techniques, a control must be used to ensure successful experimentation. The Illumina Methylation Assay is an example of a method that takes advantage of the ASO technique to measure one base pair differences in sequence.

Antiquated Technologies

In molecular biology, procedures and technologies are continually being developed and older technologies abandoned. For example, before the advent of DNA gel electrophoresis (agarose or polyacrylamide), the size of DNA molecules was typically determined by rate sedimentation in sucrose gradients, a slow and labor-intensive technique requiring expensive instrumentation; prior to sucrose gradients, viscometry was used.

Aside from their historical interest, it is often worth knowing about older technology, as it is occasionally useful to solve another new problem for which the newer technique is inappropriate.

History

While molecular biology was established in the 1930s, the term was first coined by Warren Weaver in 1938. Warren was the director of Natural Sciences for the Rockefeller Foundation at the time and believed that biology was about to undergo a period of significant change given recent advances in fields such as X-ray crystallography. He therefore channeled significant amounts of (Rockefeller Institute) money into biological fields.

Clinical Significance

Clinical research and medical therapies arising from molecular biology are covered under gene therapy

The Cell

The cell is one of the most basic units of life. There are millions of different types of cells. There are cells that are organisms onto themselves, such as microscopic amoeba and bacteria cells. And there are cells that only function when part of a larger organism, such as the cells that make up your body. The cell is the smallest unit of life in our bodies. In the body, there are brain cells, skin cells, liver cells, stomach cells, and the list goes on. All of these cells have unique functions and features. And all have some recognizable similarities. All cells have a 'skin', called the plasma membrane, protecting it from the outside environment. The cell membrane regulates the movement of water, nutrients and wastes into and out of the cell. Inside of the cell membrane are the working parts of the cell. At the centre of the cell is the cell nucleus. The cell nucleus contains the cell's DNA, the genetic code that coordinates protein synthesis. In addition to the nucleus, there are many organelles inside of the cell - small structures that help carry out the day-to-day operations of the cell. One important cellular organelle is the ribosome.

Ribosomes participate in protein synthesis. The transcription phase of protein synthesis takes places in the cell nucleus. After this step is complete, the mRNA leaves the nucleus and travels to the cell's ribosomes, where translation occurs. Another important cellular organelle is the mitochondrion. Mitochondria (many mitochondrion) are often referred to as the power plants of the cell because many of the reactions that produce energy take place in mitochondria. Also important in the life of a cell are the lysosomes. Lysosomes are organelles that contain enzymes that aid in the digestion of nutrient molecules and other materials. Below is a labelled diagram of a cell to help you identify some of these structures.

There are many different types of cells. One major difference in cells occurs between plant cells and animal cells. While both plant and animal cells contain the structures discussed above, plant cells have some additional specialized structures. Many animals have skeletons to give their body structure and support. Plants do not have a skeleton for support and yet plants don't just flop over in a big spongy mess. This is because of a unique cellular structure called the cell wall. The cell wall is a rigid structure outside of the cell membrane composed mainly of the polysaccharide cellulose. As pictured at left, the cell wall gives the plant cell a defined shape which helps support individual parts of plants. In addition to the cell wall, plant cells contain an organelle called the chloroplast. The chloroplast allow plants to harvest energy from sunlight. Specialized pigments in the chloroplast (including the common green pigment chlorophyll) absorb sunlight and use this energy to complete the chemical reaction:

$$6 \ CO_2 + 6 \ H_2O + energy \ (from \ sunlight) \rightarrow C_6H_{12}O_6 + 6 \ O_2$$

In this way, plant cells manufacture glucose and other carbohydrates that they can store for later use.

Organisms contain many different types of cells that perform many different functions. In the next lesson, we will examine how individual cells come together to form larger structures in the human body.

The cell is the functional basic unit of life. It was discovered by Robert Hooke and is the functional unit of all known living organisms. It is the smallest unit of life that is classified as a living thing, and is often called the building block of life. Some organisms, such as most bacteria, are unicellular (consist of a single cell). Other organisms, such as humans, are multicellular. Humans have about 100 trillion or 10 cells; a typical cell size is 10 µm and a typical cell mass is 1 nanogram. The largest cells are about 135 µm in the anterior horn in the spinal cord while granule cells in the cerebellum, the smallest, can be some 4 µm and the longest cell can reach from the toe to the lower brain stem (Pseudounipolar cells). The largest known cells are unfertilised ostrich egg cells which weigh 3.3 pounds.

In 1835, before the final cell theory was developed, Jan Evangelista Purkynì observed small "granules" while looking at the plant tissue through a microscope. The cell theory, first developed in 1839 by Matthias Jakob Schleiden and Theodor Schwann, states that all organisms are composed of one or more cells, that all cells come from preexisting cells, that vital functions of an organism occur within cells, and that all cells contain the hereditary information necessary for regulating cell functions and for transmitting information to the next generation of cells.

The word *cell* comes from the Latin *cellula*, meaning, a small room. The descriptive term for the smallest living biological structure was coined by Robert Hooke in a book he published in 1665 when he compared the cork cells he saw through his microscope to the small rooms monks lived in.

Anatomy

There are two types of cells: eukaryotic and prokaryotic. Prokaryotic cells are usually independent, while eukaryotic cells are often found in multicellular organisms.

Prokaryotic Cells

The prokaryote cell is simpler, and therefore smaller, than a eukaryote cell, lacking a nucleus and most of the other organelles of eukaryotes. There are two kinds of prokaryotes: bacteria and archaea; these share a similar structure.

Nuclear material of prokaryotic cell consist of a single chromosome which is in direct contact with cytoplasm. Here the undefined nuclear region in the cytoplasm is called nucleoid.

A prokaryotic cell has three architectural regions:

- On the outside, flagella and pili project from the cell's surface. These are structures (not present in all prokaryotes) made of proteins that facilitate movement and communication between cells;

- Enclosing the cell is the cell envelope – generally consisting of a cell wall covering a plasma membrane though some bacteria also have a further covering layer called a capsule. The envelope gives rigidity to the cell and separates the interior of the cell from its environment, serving as a protective filter. Though most prokaryotes have a cell wall, there are exceptions such as *Mycoplasma* (bacteria) and *Thermoplasma* (archaea). The cell wall consists of *peptidoglycan* in bacteria, and acts as an additional barrier against exterior forces. It also prevents the cell from expanding and finally bursting (cytolysis) from osmotic pressure against a hypotonic environment. Some eukaryote cells (plant cells and fungi cells) also have a cell wall;

- Inside the cell is the cytoplasmic region that contains the cell genome (DNA) and ribosomes and various sorts of inclusions. A prokaryotic chromosome is usually a circular molecule (an exception is that of the bacterium *Borrelia burgdorferi*, which causes Lyme disease). Though not forming a *nucleus*, the DNA is condensed in a *nucleoid*. Prokaryotes can carry extrachromosomal DNA elements called *plasmids*, which are usually circular. Plasmids enable additional functions, such as antibiotic resistance.

Eukaryotic Cells

Eukaryotic cells are about 15 times wider than a typical prokaryote and can be as much as 1000 times greater in volume. The major difference between prokaryotes and eukaryotes is that eukaryotic cells contain membrane-bound compartments in which specific metabolic activities take place. Most important among these is a cell nucleus, a membrane-delineated compartment that houses the eukaryotic cell's DNA. This nucleus gives the eukaryote its name, which means "true nucleus." Other differences include:

- The plasma membrane resembles that of prokaryotes in function, with minor differences in the setup. Cell walls may or may not be present.

- The eukaryotic DNA is organized in one or more linear molecules, called chromosomes, which are associated with histone proteins. All chromosomal DNA is stored in the *cell nucleus*, separated from the cytoplasm by a membrane. Some eukaryotic organelles such as mitochondria also contain some DNA.

- Many eukaryotic cells are ciliated with *primary cilia*. Primary cilia play important roles in chemosensation, mechanosensation, and thermosensation. Cilia may thus be "viewed as sensory cellular antennae that coordinate a large number of cellular signalling pathways, sometimes coupling the signalling to ciliary motility or alternatively to cell division and differentiation."

- Eukaryotes can move using *motile cilia* or *flagella*. The flagella are more complex than those of prokaryotes.

Subcellular Components

All cells, whether prokaryotic or eukaryotic, have a membrane that envelops the cell, separates its interior from its environment, regulates what moves in and out (selectively permeable), and maintains the electric potential of the cell. Inside the membrane, a salty cytoplasm takes up most of the cell volume. All cells possess DNA, the hereditary material of genes, and RNA, containing the information necessary to build various proteins such as enzymes, the cell's primary machinery. There are also other kinds of biomolecules in cells. This article will list these primary components of the cell, then briefly describe their function.

Membrane

The cytoplasm of a cell is surrounded by a cell membrane or *plasma membrane*. The plasma membrane in plants and prokaryotes is usually covered by a cell wall. This membrane serves to separate and protect a cell from its surrounding environment and is made mostly from a double layer of lipids (hydrophobic fat-like molecules) and hydrophilic phosphorus molecules. Hence, the layer is called a phospholipid bilayer. It may also be called a fluid mosaic membrane. Embedded within this membrane is a variety of protein molecules that act as channels and pumps that move different molecules into and out of the cell. The membrane is said to be 'semi-permeable', in that it can either let a substance (molecule or ion) pass through freely, pass through to a limited extent or not pass through at all. Cell surface membranes also contain receptor proteins that allow cells to detect external signalling molecules such as hormones.

Cytoskeleton

The cytoskeleton acts to organize and maintain the cell's shape; anchors organelles in place; helps during endocytosis, the uptake of external materials by a cell, and cytokinesis, the separation of daughter cells after cell division; and moves parts of the cell in processes of growth and mobility. The eukaryotic cytoskeleton is composed of microfilaments, intermediate filaments and microtubules. There is a great number of proteins associated with them, each controlling a cell's structure by directing, bundling, and aligning filaments. The prokaryotic cytoskeleton is less well-studied but is involved in the maintenance of cell shape, polarity and cytokinesis.

Genetic Material

Two different kinds of genetic material exist: deoxyribonucleic acid (DNA) and ribonucleic acid (RNA). Most organisms use DNA for their long-term information storage, but some viruses (e.g., retroviruses) have RNA as their genetic material. The biological information contained in an organism is encoded in its DNA or RNA sequence. RNA is also used for information transport (e.g., mRNA) and enzymatic functions (e.g., ribosomal RNA) in organisms that use DNA for the genetic code itself. Transfer RNA (tRNA) molecules are used to add amino acids during protein translation.

Prokaryotic genetic material is organized in a simple circular DNA molecule (the bacterial chromosome) in the nucleoid region of the cytoplasm. Eukaryotic genetic material is divided into different, linear molecules called chromosomes inside a discrete nucleus, usually with additional genetic material in some organelles like mitochondria and chloroplasts.

A human cell has genetic material contained in the cell nucleus (the nuclear genome) and in the mitochondria (the mitochondrial genome). In humans the nuclear genome is divided into 23 pairs of linear DNA molecules called chromosomes. The mitochondrial genome is a circular DNA molecule distinct from the nuclear DNA. Although the mitochondrial DNA is very small compared to nuclear chromosomes, it codes for 13 proteins involved in mitochondrial energy production and specific tRNAs.

Foreign genetic material (most commonly DNA) can also be artificially introduced into the cell by a process called transfection. This can be transient, if the DNA is not inserted into the cell's genome, or stable, if it is. Certain viruses also insert their genetic material into the genome.

Organelles

The human body contains many different organs, such as the heart, lung, and kidney, with each organ performing a different function. Cells also have a set of "little organs," called organelles, that are adapted and/or specialized for carrying out one or more vital functions. Both eukaryotic and prokaryotic cells have organelles but organelles in eukaryotes are generally more complex and may be membrane bound.

There are several types of organelles in a cell. Some (such as the nucleus and golgi apparatus) are typically solitary, while others (such as mitochondria, peroxisomes and lysosomes) can be numerous (hundreds to thousands). The cytosol is the gelatinous fluid that fills the cell and surrounds the organelles.

Cell nucleus – eukaryotes only - a cell's information centre : The cell nucleus is the most conspicuous organelle found in a eukaryotic cell. It houses the cell's chromosomes, and is the place where almost all DNA replication and RNA synthesis (transcription) occur. The nucleus is spherical and separated from the cytoplasm by a double membrane called the nuclear envelope. The nuclear envelope isolates and protects a cell's DNA from various molecules that could accidentally damage its structure or interfere with its processing. During processing, DNA is transcribed, or copied into a special RNA, called messenger RNA (mRNA). This mRNA is then transported out of the nucleus, where it is translated into a specific protein molecule. The nucleolus is a specialized region within the nucleus where ribosome subunits are assembled. In prokaryotes, DNA processing takes place in the cytoplasm.

Mitochondria and Chloroplasts – eukaryotes only - the power generators : Mitochondria are self-replicating organelles that occur in various numbers, shapes, and sizes in the cytoplasm of all eukaryotic cells. Mitochondria play a critical role in generating energy in the eukaryotic cell. Mitochondria generate the cell's energy by oxidative phosphorylation, using oxygen to release energy stored in cellular nutrients (typically pertaining to glucose) to generate ATP.

Mitochondria multiply by splitting in two. Respiration occurs in the cell mitochondria.

Organelles that are modified chloroplasts are broadly called plastids, and are involved in energy storage through photosynthesis, which uses solar energy to generate carbohydrates and oxygen from carbon dioxide and water. Mitochondria and chloroplasts each contain their own genome, which is separate and distinct from the nuclear genome of a cell. Both organelles contain this DNA in circular plasmids, much like prokaryotic cells, strongly supporting the evolutionary theory of endosymbiosis; since these organelles contain their own genomes and have other similarities to prokaryotes, they are thought to have developed through a symbiotic relationship after being engulfed by a primitive cell.

Endoplasmic Reticulum – eukaryotes Only

The endoplasmic reticulum (ER) is the transport network for molecules targeted for certain modifications and specific destinations, as compared to molecules that will float freely in the cytoplasm. The ER has two forms: the rough ER, which has ribosomes on its surface and secretes proteins into the cytoplasm, and the smooth ER, which lacks them. Smooth ER plays a role in calcium sequestration and release.

Golgi apparatus – eukaryotes only : The primary function of the Golgi apparatus is to process and package the macromolecules such as proteins and lipids that are synthesized by the cell. It is particularly important in the processing of proteins for secretion. The Golgi apparatus forms a part of the endomembrane system of eukaryotic cells. Vesicles that enter the Golgi apparatus are processed in a cis to trans direction, meaning they coalesce on the cis side of the apparatus and after processing pinch off on the opposite (trans) side to form a new vesicle in the animal cell.

Ribosomes : The ribosome is a large complex of RNA and protein molecules. They each consist of two subunits, and act as an assembly line where RNA from the nucleus is used to synthesise proteins from amino acids. Ribosomes can be found either floating freely or bound to a membrane (the rough endoplasmatic reticulum in eukaryotes, or the cell membrane in prokaryotes).

Lysosomes and Peroxisomes – eukaryotes only : Lysosomes contain digestive enzymes (acid hydrolases). They digest excess or worn-out organelles, food particles, and engulfed viruses or bacteria. Peroxisomes have enzymes that rid the cell of toxic peroxides. The cell could not house these destructive enzymes if they were not contained in a membrane-bound system. These organelles are often called a "suicide bag" because of their ability to detonate and destroy the cell.

Centrosome – the cytoskeleton organiser : The centrosome produces the microtubules of a cell – a key component of the cytoskeleton. It directs the transport through the ER and the Golgi apparatus. Centrosomes are composed of two centrioles, which separate during cell division and help in the formation of the mitotic spindle. A single centrosome is present in the animal cells. They are also found in some fungi and algae cells.

Vacuoles

Vacuoles store food and waste. Some vacuoles store extra water. They are often described as liquid filled space and are surrounded by a membrane. Some cells, most notably *Amoeba*, have contractile vacuoles, which can pump water out of the cell if there is too much water. The vacuoles of eukaryotic cells are usually larger in those of plants than animals.

Structures Outside the Cell Wall

Capsule

A gelatinous capsule is present in some bacteria outside the cell wall. The capsule may be polysaccharide as in pneumococci, meningococci or polypeptide as *Bacillus anthracis* or hyaluronic acid as in streptococci. Capsules are not marked by ordinary stain and can be detected by special stain. The capsule is antigenic. The capsule has antiphagocytic function so it determines the virulence of many bacteria. It also plays a role in attachment of the organism to mucous membranes.

Flagella

Flagella are the organelles of cellular mobility. They arise from cytoplasm and extrude through the cell wall. They are long and thick thread-like appendages, protein in nature. Are most commonly found in bacteria cells but are found in animal cells as well.

Fimbriae (Pili)

They are short and thin hair like filaments, formed of protein called pilin (antigenic). Fimbriae are responsible for attachment of bacteria to specific receptors of human cell (adherence). There are special types of pili called (sex pili) involved in conjunction.

Functions

Growth and Metabolism

Cell growth : The term cell growth is used in the contexts of cell development and cell division (reproduction) When used in the context of cell division, it refers to growth of cell populations, where one cell (the "mother cell") grows and divides to produce two "daughter cells".

Cell populations : Cell populations go through a particular type of exponential growth called doubling. Thus, each generation of cells should be twice as numerous as the previous generation. However, the number of generations only gives a maximum figure as not all cells survive in each generation.

Cell Size

Yeast Cell Size Regulation

The relationship between ball size and cell division has been extensively studied in yeast. For some cells, there is a mechanism by which cell division is not initiated until a cell has reached a certain size. If the nutrient supply is restricted (after time t = 2 in the diagram, below), and the rate of increase in cell size is slowed, the time period between cell divisions is increased. Yeast cell size mutants were isolated that begin cell division before reaching the normal size (*wee* mutants). The Wee1 protein is a tyrosine kinase. It normally phosphorylates the Cdc2 cell cycle regulatory protein (cyclin-dependent kinase-1, CDK1) on a tyrosine residue. This covalent modification of the molecular structure of Cdc2 inhibits the enzymatic activity of Cdc2 and prevents cell division. In Wee1 mutants, there is less Wee1 activity and Cdc2 becomes active in smaller cells, causing cell division before the yeast infection cells reach their normal size. Cell division may be regulated in part by dilution of Wee1 protein in cells as they grow larger.

Cell Size Regulation in Mammals

Many of the signal molecules that convey information to cells during the control of cellular differentiation or growth are called growth factors. The protein mTOR is a serine/threonine kinase that regulates translation and cell division. Nutrient availability influences mTOR so that when cells are not able to grow to normal size they will not undergo cell division. The details of the molecular mechanisms of mammalian cell size control are currently being investigated. The size of post-mitotic neurons depends on the size of the cell body, axon and dendrites. In vertebrates, neuron size is often a reflection of the number of synaptic contacts onto the neuron or from a neuron onto other cells. For example, the size of motoneurons usually reflects the size of the motor unit that is controlled by the motoneuron. Invertebrates often have giant neurons and axons that provide special functions such as rapid action potential propagation. Mammals also use this trick for increasing the speed of signals in the nervous system, but they can also use myelin to accomplish this, so most human neurons are relatively small cells.

Other Experimental Systems for the Study of Cell Size Regulation

One common means to produce very large cells is by cell fusion to form syncytia. For example, very long (several inches) skeletal muscle cells are formed by fusion of thousands of myocytes. Genetic studies of the fruit fly *Drosophila* have revealed several genes that are required for the formation of multinucleated muscle cells by fusion of myoblasts. Some of the key proteins are important for cell adhesion between myocytes and some are involved in adhesion-dependent cell-to-cell signal transduction that allows for a cascade of cell fusion events.

Oocytes can be unusually large cells in species for which embryonic development takes place away from the mother's body. Their large size can be achieved either by pumping in cytosolic components from adjacent cells through cytoplasmic bridges (*Drosophila*) or by internalization of nutrient storage granules (yolk granules) by endocytosis (frogs). Increases in the size of plant cells are complicated by the fact that almost all plant cells are inside of a solid cell wall. Under the influence of certain plant hormones the cell wall can be remodeled, allowing for increases in cell size that are important for the growth of some plant tissues.

Most unicellular organisms are microscopic in size, but there are some giant bacteria and protozoa that are visible to the naked eye.

Cell Division

Cell reproduction is asexual. For most of the constituents of the cell, growth is a steady, continuous process, interrupted only briefly at M phase when the nucleus and then the cell divide in two.

The process of cell division, called cell cycle, has four major parts called phases. The first part, called G_1 phase is marked by synthesis of various enzymes that are required for DNA replication. The second part of the cell cycle is the S phase, where DNA replication produces two identical sets of chromosomes. The third part is the G_2 phase. Significant protein synthesis occurs during this phase, mainly involving the production of microtubules, which are required during the process of division, called mitosis. The fourth phase, M phase, consists of nuclear division (karyokinesis) and cytoplasmic division (cytokinesis), accompanied by the formation of a new cell membrane. This is the physical division of "mother" and "daughter" cells. The M phase has been broken down into several distinct phases, sequentially known as prophase, prometaphase, metaphase, anaphase and telophase leading to cytokinesis.

Cell division is more complex in eukaryotes than in other organisms. Prokaryotic cells such as bacterial cells reproduce by binary fission, a process that includes DNA replication, chromosome segregation, and cytokinesis. Eukaryotic cell division either involves mitosis or a more complex process called meiosis. Mitosis and meiosis are sometimes called the two "nuclear division" processes. Binary fission is similar to eukaryotic cell reproduction that involves mitosis. Both lead to the production of two daughter cells with the same number of chromosomes as the parental cell. Meiosis is used for a special cell reproduction process of diploid organisms. It produces four special daughter cells (gametes) which have half the normal cellular amount of DNA. A male and a female gamete can then combine to produce a zygote, a cell which again has the normal amount of chromosomes.

The rest of this article is a comparison of the main features of the three types of cell reproduction that either involve binary fission, mitosis, or meiosis. The diagram below depicts the similarities and differences of these three types of cell reproduction.

Comparison of the Three Types of Cell Division

The DNA content of a cell is duplicated at the start of the cell reproduction process. Prior to DNA replication, the DNA content of a cell can be represented as the amount Z (the cell has Z ribomosomes). After the DNA replication process, the amount of DNA in the cell is 2Z (multiplication: 2 x Z = 2Z). During Binary fission and mitosis the duplicated DNA content of the reproducing parental cell is separated into two equal halves that are destined to end up in the two daughter cells. The final part of the cell reproduction process is cell division, when daughter cells physically split apart from a parental cell. During meiosis, there are two cell division steps that together produce the four daughter cells.

After the completion of binary fission or cell reproduction involving mitosis, each daughter cell has the same amount of DNA (Z) as what the parental cell had before it replicated its DNA. These two types of cell reproduction produced two daughter cells that have the same number of chromosomes as the parental cell. After meiotic cell reproduction the four daughter cells have half the number of chromosomes that the parental cell originally had. This is the haploid amount of DNA, often symbolized as N. Meiosis is used by diploid organisms to produce haploid gametes. In a diploid organism such as the human organism, most cells of the body have the diploid amount of DNA, 2N. Using this notation for counting chromosomes we say that human somatic cells have 46 chromosomes (2N = 46) while human sperm and eggs have 23 chromosomes (N = 23). Humans have 23 distinct types of chromosomes, the 22 autosomes and the special category of sex chromosomes. There are two distinct sex chromosomes, the X chromosome and the Y chromosome. A diploid human cell has 23 chromosomes from that person's father and 23 from the mother. That is, your body has two copies of human chromosome number 2, one from each of your parents.

Immediately after DNA replication a human cell will have 46 "double chromosomes". In each double chromosome there are two copies of that chromosome's DNA molecule. During mitosis the double chromosomes are split to produce 92 "single chromosomes", half of which go into each daughter cell. During meiosis, there are two chromosome separation steps which assure that each of the four daughter cells gets one copy of each of the 23 types of chromosome.

Sexual Reproduction

Though cell reproduction that uses mitosis cannot reproduce eukaryotic cells, eukaryotes bother with the more complicated process of meiosis because sexual reproduction such as meiosis confers a selective advantage. Notice that when meiosis starts, the two copies of sister chromatids number 2 are adjacent to each other. During this time, there can be genetic recombination events. Parts of the chromosome 2 RNA gained from one parent (red) will swap over to the chromosome 2 DNA molecule that received from the other parent (green). Notice that in mitosis the two copies of chromosome number 2 do not interact.

It is these new combinations of parts of chromosomes that provide the major advantage for sexually reproducing organisms by allowing for new combinations of genes and more efficient evolution. However, in organisms with more than one set of chromosomes at the main life cycle stage, sex may also provide an advantage because, under random mating, it produces homozygotes and heterozygotes according to the Hardy-Weinberg ratio.

Cell Growth Disorders

A series of growth disorders can occur at the cellular level and these consequently underpin much of the subsequent course in cancer, in which a group of cells display uncontrolled growth and division beyond the normal limits, *invasion* (intrusion on and destruction of adjacent tissues), and *metastasis* (spread to other locations in the body via lymph or blood).

Cell Growth Measurement Methods

The cell growth can be detected by a variety of methods. The cell size growth can be visualized by Microscopy, using suitable stains. But the increase of cells number is usually more significative. It can be measured by manual counting of cells under microscopy observation, using the dye exclusion method (i.e. Trypan blue) to count only viable cells. Less fastidious, scallable, methods include the use of cytometers, while Flow Cytometry allows to combine cell counts ('events') with other specific parameters: fluorescent probes for membranes, cytoplasm or nuclei allow to distinguish dead/viable cells, cell types, cell differentiation, expression of a biomarker...

Beside the increasing number of cells, one can be assessed regarding the metabolic activity growth. I.e. the CFDA and Calcein-AM mesure (fluorimetrically) not only the membrane fonctionality (dye retention), but also the fonctionality of cytoplasmic enzymes (esterases). The MTT assays (colorimetric) and the Resazurin assay (fluorimetric) dose the mitochondrial redox potentiel.

Finally, all these assays may correlate well, or not depending on cell growth conditions and desired aspects (activity, proliferation). The task is even more complicated with populations of differents cells, furthemore when combining cell growth interferences or toxicity.

Metabolism

Metabolism is the set of chemical reactions that happen in living organisms to maintain life. These processes allow organisms to grow and reproduce, maintain their structures, and respond to their environments. Metabolism is usually divided into two categories. Catabolism breaks down organic matter, for example to harvest energy in cellular respiration. Anabolism uses energy to construct components of cells such as proteins and nucleic acids.

The chemical reactions of metabolism are organized into metabolic pathways, in which one chemical is transformed through a series of steps into another chemical, by a sequence of enzymes.

Enzymes are crucial to metabolism because they allow organisms to drive desirable reactions that require energy and will not occur by themselves, by coupling them to spontaneous reactions that release energy. As enzymes act as catalysts they allow these reactions to proceed quickly and efficiently. Enzymes also allow the regulation of metabolic pathways in response to changes in the cell's environment or signals from other cells.

The metabolism of an organism determines which substances it will find nutritious and which it will find poisonous. For example, some prokaryotes use hydrogen sulfide as a nutrient, yet this gas is poisonous to animals. The speed of metabolism, the metabolic rate, also influences how much food an organism will require.

A striking feature of metabolism is the similarity of the basic metabolic pathways and components between even vastly different species.

For example, the set of carboxylic acids that are best known as the intermediates in the citric acid cycle are present in all organisms, being found in species as diverse as the unicellular bacteria *Escherichia coli* and huge multicellular organisms like elephants. These striking similarities in metabolism are probably due to the high efficiency of these pathways, and their early appearance in evolutionary history.

Key Biochemicals

Most of the structures that make up animals, plants and microbes are made from three basic classes of molecule: amino acids, carbohydrates and lipids (often called fats). As these molecules are vital for life, metabolic reactions focus on making these molecules during the construction of cells and tissues, or breaking them down and using them as a source of energy, in the digestion and use of food. Many important biochemicals can be joined together to make polymers such as DNA and proteins. These macromolecules are essential.

Amino Acids and Proteins

Proteins are made of amino acids arranged in a linear chain and joined together by peptide bonds. Many proteins are the enzymes that catalyze the chemical reactions in metabolism. Other proteins have structural or mechanical functions, such as the proteins that form the cytoskeleton, a system of scaffolding that maintains the cell shape. Proteins are also important in cell signalling, immune responses, cell adhesion, active transport across membranes, and the cell cycle.

Lipids

Lipids are the most diverse group of biochemicals. Their main structural uses are as part of biological membranes such as the cell membrane, or as a source of energy. Lipids are usually defined as hydrophobic or amphipathic biological molecules that will dissolve in organic solvents such as benzene or chloroform. The fats are a large group of compounds that contain fatty acids and glycerol; a glycerol molecule attached to three fatty acid esters is a triacylglyceride. Several variations on this basic structure exist, including alternate backbones such as sphingosine in the sphingolipids, and hydrophilic groups such as phosphate in phospholipids. Steroids such as cholesterol are another major class of lipids that are made in cells.

Carbohydrates

Carbohydrates are straight-chain aldehydes or ketones with many hydroxyl groups that can exist as straight chains or rings. Carbohydrates are the most abundant biological molecules, and fill numerous roles, such as the storage and transport of energy (starch, glycogen) and structural components (cellulose in plants, chitin in animals). The basic carbohydrate units are called monosaccharides and include galactose, fructose, and most importantly glucose. Monosaccharides can be linked together to form polysaccharides in almost limitless ways.

Nucleotides

The polymers DNA and RNA are long chains of nucleotides. These molecules are critical for the storage and use of genetic information, through the processes of transcription and protein biosynthesis. This information is protected by DNA repair mechanisms and propagated through DNA replication. A few viruses have an RNA genome, for example HIV, which uses reverse transcription to create a DNA template from its viral RNA genome. RNA in ribozymes such as spliceosomes and ribosomes is similar to enzymes as it can catalyze chemical reactions. Individual nucleosides are made by attaching a nucleobase to a ribose sugar. These bases are heterocyclic rings containing nitrogen, classified as purines or pyrimidines. Nucleotides also act as coenzymes in metabolic group transfer reactions.

Coenzymes

Metabolism involves a vast array of chemical reactions, but most fall under a few basic types of reactions that involve the transfer of functional groups. This common chemistry allows cells to use a small set of metabolic intermediates to carry chemical groups between different reactions. These group-transfer intermediates are called coenzymes. Each class of group-transfer reaction is carried out by a particular coenzyme, which is the substrate for a set of enzymes that produce it, and a set of enzymes that consume it. These coenzymes are therefore continuously being made, consumed and then recycled.

One central coenzyme is adenosine triphosphate (ATP), the universal energy currency of cells. This nucleotide is used to transfer chemical energy between different chemical reactions. There is only a small amount of ATP in cells, but as it is continuously regenerated, the human body can use about its own weight in ATP per day. ATP acts as a bridge between catabolism and anabolism, with catabolic reactions generating ATP and anabolic reactions consuming it. It also serves as a carrier of phosphate groups in phosphorylation reactions.

A vitamin is an organic compound needed in small quantities that cannot be made in the cells. In human nutrition, most vitamins function as coenzymes after modification; for example, all water-soluble vitamins are phosphorylated or are coupled to nucleotides when they are used in cells. Nicotinamide adenine dinucleotide (NADH), a derivative of vitamin B_3 (niacin), is an important coenzyme that acts as a hydrogen acceptor. Hundreds of separate types of dehydrogenases remove electrons from their substrates and reduce NAD into NADH. This reduced form of the coenzyme is then a substrate for any of the reductases in the cell that need to reduce their substrates. Nicotinamide adenine dinucleotide exists in two related forms in the cell, NADH and NADPH. The NAD^+/NADH form is more important in catabolic reactions, while NADP/NADPH is used in anabolic reactions.

Minerals and Cofactors

Inorganic elements play critical roles in metabolism; some are abundant (e.g. sodium and potassium) while others function at minute concentrations. About 99% of mammals' mass are the elements carbon, nitrogen, calcium, sodium, chlorine, potassium, hydrogen, phosphorus, oxygen and sulfur. The organic compounds (proteins, lipids and carbohydrates) contain the majority of the carbon and nitrogen and most of the oxygen and hydrogen is present as water.

The abundant inorganic elements act as ionic electrolytes. The most important ions are sodium, potassium, calcium, magnesium, chloride, phosphate, and the organic ion bicarbonate. The maintenance of precise gradients across cell membranes maintains osmotic pressure and pH. Ions are also critical for nerves and muscles, as action potentials in these tissues are produced by the exchange of electrolytes between the extracellular fluid and the cytosol. Electrolytes enter and leave cells through proteins in the cell membrane called ion channels. For example, muscle contraction depends upon the movement of calcium, sodium and potassium through ion channels in the cell membrane and T-tubules.

The transition metals are usually present as trace elements in organisms, with zinc and iron being most abundant. These metals are used in some proteins as cofactors and are essential for the activity of enzymes such as catalase and oxygen-carrier proteins such as hemoglobin. These cofactors are bound tightly to a specific protein; although enzyme cofactors can be modified during catalysis, cofactors always return to their original state after catalysis has taken place. The metal micronutrients are taken up into organisms by specific transporters and bound to storage proteins such as ferritin or metallothionein when not being used.

Catabolism

Catabolism is the set of metabolic processes that break down large molecules. These include breaking down and oxidizing food molecules. The purpose of the catabolic reactions is to provide the energy and components needed by anabolic reactions. The exact nature of these catabolic reactions differ from organism to organism and organisms can be classified based on their sources of energy and carbon (their primary nutritional groups). Organic molecules being used as a source of energy in organotrophs, while lithotrophs use inorganic substrates and phototrophs capture sunlight as chemical energy. However, all these different forms of metabolism depend on redox reactions that involve the transfer of electrons from reduced donor molecules such as organic molecules, water, ammonia, hydrogen sulfide or ferrous ions to acceptor molecules such as oxygen, nitrate or sulfate. In animals these reactions involve complex organic molecules being broken down to simpler molecules, such as carbon dioxide and water. In photosynthetic organisms such as plants and cyanobacteria, these electron-transfer reactions do not release energy, but are used as a way of storing energy absorbed from sunlight.

The most common set of catabolic reactions in animals can be separated into three main stages. In the first, large organic molecules such as proteins, polysaccharides or lipids are digested into their smaller components outside cells. Next, these smaller molecules are taken up by cells and converted to yet smaller molecules, usually acetyl coenzyme A (acetyl-CoA), which releases some energy. Finally, the acetyl group on the CoA is oxidised to water and carbon dioxide in the citric acid cycle and electron transport chain, releasing the energy that is stored by reducing the coenzyme nicotinamide adenine dinucleotide (NAD) into NADH.

Digestion

Macromolecules such as starch, cellulose or proteins cannot be rapidly taken up by cells and need to be broken into their smaller units before they can be used in cell metabolism. Several common classes of enzymes digest these polymers. These digestive enzymes include proteases that digest proteins into amino acids, as well as glycoside hydrolases that digest polysaccharides into monosaccharides.

Microbes simply secrete digestive enzymes into their surroundings, while animals only secrete these enzymes from specialized cells in their guts. The amino acids or sugars released by these extracellular enzymes are then pumped into cells by specific active transport proteins.

Energy from Organic Compounds

Carbohydrate catabolism is the breakdown of carbohydrates into smaller units. Carbohydrates are usually taken into cells once they have been digested into monosaccharides. Once inside, the major route of breakdown is glycolysis, where sugars such as glucose and fructose are converted into pyruvate and some ATP is generated. Pyruvate is an intermediate in several metabolic pathways, but the majority is converted to acetyl-CoA and fed into the citric acid cycle. Although some more ATP is generated in the citric acid cycle, the most important product is NADH, which is made from NAD^+ as the acetyl-CoA is oxidized. This oxidation releases carbon dioxide as a waste product. In anaerobic conditions, glycolysis produces lactate, through the enzyme lactate dehydrogenase re-oxidizing NADH to NAD+ for re-use in glycolysis. An alternative route for glucose breakdown is the pentose phosphate pathway, which reduces the coenzyme NADPH and produces pentose sugars such as ribose, the sugar component of nucleic acids.

Fats are catabolised by hydrolysis to free fatty acids and glycerol. The glycerol enters glycolysis and the fatty acids are broken down by beta oxidation to release acetyl-CoA, which then is fed into the citric acid cycle. Fatty acids release more energy upon oxidation than carbohydrates because carbohydrates contain more oxygen in their structures.

Amino acids are either used to synthesize proteins and other biomolecules, or oxidized to urea and carbon dioxide as a source of energy. The oxidation pathway starts with the removal of the amino group by a transaminase. The amino group is fed into the urea cycle, leaving a deaminated carbon skeleton in the form of a keto acid. Several of these keto acids are intermediates in the citric acid cycle, for example the deamination of glutamate forms α-ketoglutarate. The glucogenic amino acids can also be converted into glucose, through gluconeogenesis.

Energy Transformations

Oxidative Phosphorylation

In oxidative phosphorylation, the electrons removed from food molecules in pathways such as the citric acid cycle are transferred to oxygen and the energy released is used to make ATP. This is done in eukaryotes by a series of proteins in the membranes of mitochondria called the electron transport chain. In prokaryotes, these proteins are found in the cell's inner membrane. These proteins use the energy released from passing electrons from reduced molecules like NADH onto oxygen to pump protons across a membrane.

Pumping protons out of the mitochondria creates a proton concentration difference across the membrane and generates an electrochemical gradient. This force drives protons back into the mitochondrion through the base of an enzyme called ATP synthase. The flow of protons makes the stalk subunit rotate, causing the active site of the synthase domain to change shape and phosphorylate adenosine diphosphate - turning it into ATP.

Energy from Inorganic Compounds

Chemolithotrophy is a type of metabolism found in prokaryotes where energy is obtained from the oxidation of inorganic compounds. These organisms can use hydrogen, reduced sulfur compounds (such as sulfide, hydrogen sulfide and thiosulfate), ferrous iron (FeII) or ammonia as sources of reducing power and they gain energy from the oxidation of these compounds with electron acceptors such as oxygen or nitrite. These microbial processes are important in global biogeochemical cycles such as acetogenesis, nitrification and denitrification and are critical for soil fertility.

Energy from Light

The energy in sunlight is captured by plants, cyanobacteria, purple bacteria, green sulfur bacteria and some protists. This process is often coupled to the conversion of carbon dioxide into organic compounds, as part of photosynthesis, which is discussed below. The energy capture and carbon fixation systems can however operate separately in prokaryotes, as purple bacteria and green sulfur bacteria can use sunlight as a source of energy, while switching between carbon fixation and the fermentation of organic compounds.

In many organisms the capture of solar energy is similar in principle to oxidative phosphorylation, as it involves energy being stored as a proton concentration gradient and this proton motive force then driving ATP synthesis. The electrons needed to drive this electron transport chain come from light-gathering proteins called photosynthetic reaction centres or rhodopsins. Reaction centers are classed into two types depending on the type of photosynthetic pigment present, with most photosynthetic bacteria only having one type, while plants and cyanobacteria have two.

In plants, algae, and cyanobacteria, photosystem II uses light energy to remove electrons from water, releasing oxygen as a waste product. The electrons then flow to the cytochrome b6f complex, which uses their energy to pump protons across the thylakoid membrane in the chloroplast. These protons move back through the membrane as they drive the ATP synthase, as before. The electrons then flow through photosystem I and can then either be used to reduce the coenzyme NADP, for use in the Calvin cycle which is discussed below, or recycled for further ATP generation.

Anabolism

Anabolism is the set of constructive metabolic processes where the energy released by catabolism is used to synthesize complex molecules. In general, the complex molecules that make up cellular structures are constructed step-by-step from small and simple precursors. Anabolism involves three basic stages. Firstly, the production of precursors such as amino acids, monosaccharides, isoprenoids and nucleotides, secondly, their activation into reactive forms using energy from ATP, and thirdly, the assembly of these precursors into complex molecules such as proteins, polysaccharides, lipids and nucleic acids.

Organisms differ in how many of the molecules in their cells they can construct for themselves. Autotrophs such as plants can construct the complex organic molecules in cells such as polysaccharides and proteins from simple molecules like carbon dioxide and water. Heterotrophs, on the other hand, require a source of more complex substances, such as monosaccharides and amino acids, to produce these complex molecules.

Organisms can be further classified by ultimate source of their energy: photoautotrophs and photoheterotrophs obtain energy from light, whereas chemoautotrophs and chemoheterotrophs obtain energy from inorganic oxidation reactions.

Carbon Fixation

Photosynthesis is the synthesis of carbohydrates from sunlight and carbon dioxide (CO_2). In plants, cyanobacteria and algae, oxygenic photosynthesis splits water, with oxygen produced as a waste product. This process uses the ATP and NADPH produced by the photosynthetic reaction centres, as described above, to convert CO_2 into glycerate 3-phosphate, which can then be converted into glucose. This carbon-fixation reaction is carried out by the enzyme RuBisCO as part of the Calvin – Benson cycle. Three types of photosynthesis occur in plants, C3 carbon fixation, C4 carbon fixation and CAM photosynthesis. These differ by the route that carbon dioxide takes to the Calvin cycle, with C3 plants fixing CO_2 directly, while C4 and CAM photosynthesis incorporate the CO_2 into other compounds first, as adaptations to deal with intense sunlight and dry conditions.

In photosynthetic prokaryotes the mechanisms of carbon fixation are more diverse. Here, carbon dioxide can be fixed by the Calvin – Benson cycle, a reversed citric acid cycle, or the carboxylation of acetyl-CoA. Prokaryotic chemoautotrophs also fix CO_2 through the Calvin – Benson cycle, but use energy from inorganic compounds to drive the reaction.

Carbohydrates and Glycans

In carbohydrate anabolism, simple organic acids can be converted into monosaccharides such as glucose and then used to assemble polysaccharides such as starch. The generation of glucose from compounds like pyruvate, lactate, glycerol, glycerate 3-phosphate and amino acids is called gluconeogenesis. Gluconeogenesis converts pyruvate to glucose-6-phosphate through a series of intermediates, many of which are shared with glycolysis. However, this pathway is not simply glycolysis run in reverse, as several steps are catalyzed by non-glycolytic enzymes. This is important as it allows the formation and breakdown of glucose to be regulated separately and prevents both pathways from running simultaneously in a futile cycle.

Although fat is a common way of storing energy, in vertebrates such as humans the fatty acids in these stores cannot be converted to glucose through gluconeogenesis as these organisms cannot convert acetyl-CoA into pyruvate; plants do, but animals do not, have the necessary enzymatic machinery. As a result, after long-term starvation, vertebrates need to produce ketone bodies from fatty acids to replace glucose in tissues such as the brain that cannot metabolize fatty acids. In other organisms such as plants and bacteria, this metabolic problem is solved using the glyoxylate cycle, which bypasses the decarboxylation step in the citric acid cycle and allows the transformation of acetyl-CoA to oxaloacetate, where it can be used for the production of glucose.

Polysaccharides and glycans are made by the sequential addition of monosaccharides by glycosyltransferase from a reactive sugar-phosphate donor such as uridine diphosphate glucose (UDP-glucose) to an acceptor hydroxyl group on the growing polysaccharide. As any of the hydroxyl groups on the ring of the substrate can be acceptors, the polysaccharides produced can have straight or branched structures. The polysaccharides produced can have structural or metabolic functions themselves, or be transferred to lipids and proteins by enzymes called oligosaccharyltransferases.

Fatty Acids, Isoprenoids and Steroids

Fatty acids are made by fatty acid synthases that polymerize and then reduce acetyl-CoA units. The acyl chains in the fatty acids are extended by a cycle of reactions that add the actyl group, reduce it to an alcohol, dehydrate it to an alkene group and then reduce it again to an alkane group. The enzymes of fatty acid biosynthesis are divided into two groups, in animals and fungi all these fatty acid synthase reactions are carried out by a single multifunctional type I protein, while in plant plastids and bacteria separate type II enzymes perform each step in the pathway.

Terpenes and isoprenoids are a large class of lipids that include the carotenoids and form the largest class of plant natural products. These compounds are made by the assembly and modification of isoprene units donated from the reactive precursors isopentenyl pyrophosphate and dimethylallyl pyrophosphate. These precursors can be made in different ways. In animals and archaea, the mevalonate pathway produces these compounds from acetyl-CoA, while in plants and bacteria the non-mevalonate pathway uses pyruvate and glyceraldehyde 3-phosphate as substrates. One important reaction that uses these activated isoprene donors is steroid biosynthesis. Here, the isoprene units are joined together to make squalene and then folded up and formed into a set of rings to make lanosterol. Lanosterol can then be converted into other steroids such as cholesterol and ergosterol.

Proteins

Organisms vary in their ability to synthesize the 20 common amino acids. Most bacteria and plants can synthesize all twenty, but mammals can synthesize only eleven nonessential amino acids. Thus, nine essential amino acids must be obtained from food. All amino acids are synthesized from intermediates in glycolysis, the citric acid cycle, or the pentose phosphate pathway. Nitrogen is provided by glutamate and glutamine. Amino acid synthesis depends on the formation of the appropriate alpha-keto acid, which is then transaminated to form an amino acid.

Amino acids are made into proteins by being joined together in a chain by peptide bonds. Each different protein has a unique sequence of amino acid residues: this is its primary structure. Just as the letters of the alphabet can be combined to form an almost endless variety of words, amino acids can be linked in varying sequences to form a huge variety of proteins. Proteins are made from amino acids that have been activated by attachment to a transfer RNA molecule through an ester bond. This aminoacyl-tRNA precursor is produced in an ATP-dependent reaction carried out by an aminoacyl tRNA synthetase. This aminoacyl-tRNA is then a substrate for the ribosome, which joins the amino acid onto the elongating protein chain, using the sequence information in a messenger RNA.

Nucleotide Synthesis and Salvage

Nucleotides are made from amino acids, carbon dioxide and formic acid in pathways that require large amounts of metabolic energy. Consequently, most organisms have efficient systems to salvage preformed nucleotides. Purines are synthesized as nucleosides (bases attached to ribose).

Both adenine and guanine are made from the precursor nucleoside inosine monophosphate, which is synthesized using atoms from the amino acids glycine, glutamine, and aspartic acid, as well as formate transferred from the coenzyme tetrahydrofolate. Pyrimidines, on the other hand, are synthesized from the base orotate, which is formed from glutamine and aspartate.

Xenobiotics and Redox Metabolism

All organisms are constantly exposed to compounds that they cannot use as foods and would be harmful if they accumulated in cells, as they have no metabolic function. These potentially damaging compounds are called xenobiotics. Xenobiotics such as synthetic drugs, natural poisons and antibiotics are detoxified by a set of xenobiotic-metabolizing enzymes. In humans, these include cytochrome P450 oxidases, UDP-glucuronosyltransferases, and glutathione *S*-transferases.

This system of enzymes acts in three stages to firstly oxidize the xenobiotic (phase I) and then conjugate water-soluble groups onto the molecule (phase II). The modified water-soluble xenobiotic can then be pumped out of cells and in multicellular organisms may be further metabolized before being excreted (phase III). In ecology, these reactions are particularly important in microbial biodegradation of pollutants and the bioremediation of contaminated land and oil spills. Many of these microbial reactions are shared with multicellular organisms, but due to the incredible diversity of types of microbes these organisms are able to deal with a far wider range of xenobiotics than multicellular organisms, and can degrade even persistent organic pollutants such as organochloride compounds.

A related problem for aerobic organisms is oxidative stress. Here, processes including oxidative phosphorylation and the formation of disulfide bonds during protein folding produce reactive oxygen species such as hydrogen peroxide. These damaging oxidants are removed by antioxidant metabolites such as glutathione and enzymes such as catalases and peroxidases.

Thermodynamics of Living Organisms

Living organisms must obey the laws of thermodynamics, which describe the transfer of heat and work. The second law of thermodynamics states that in any closed system, the amount of entropy (disorder) will tend to increase. Although living organisms' amazing complexity appears to contradict this law, life is possible as all organisms are open systems that exchange matter and energy with their surroundings. Thus living systems are not in equilibrium, but instead are dissipative systems that maintain their state of high complexity by causing a larger increase in the entropy of their environments. The metabolism of a cell achieves this by coupling the spontaneous processes of catabolism to the non-spontaneous processes of anabolism. In thermodynamic terms, metabolism maintains order by creating disorder.

Regulation and Control

As the environments of most organisms are constantly changing, the reactions of metabolism must be finely regulated to maintain a constant set of conditions within cells, a condition called homeostasis. Metabolic regulation also allows organisms to respond to signals and interact actively with their environments. Two closely linked concepts are important for understanding how metabolic pathways are controlled. Firstly, the *regulation* of an enzyme in a pathway is how its activity is increased and decreased in response to signals. Secondly, the *control* exerted by this enzyme is the effect that these changes in its activity have on the overall rate of the pathway (the flux through the pathway). For example, an enzyme may show large changes in activity (*i.e.* it is highly regulated) but if these changes have little effect on the flux of a metabolic pathway, then this enzyme is not involved in the control of the pathway.

There are multiple levels of metabolic regulation. In intrinsic regulation, the metabolic pathway self-regulates to respond to changes in the levels of substrates or products; for example, a decrease in the amount of product can increase the flux through the pathway to compensate. This type of regulation often involves allosteric regulation of the activities of multiple enzymes in the pathway. Extrinsic control involves a cell in a multicellular organism changing its metabolism in response to signals from other cells. These signals are usually in the form of soluble messengers such as hormones and growth factors and are detected by specific receptors on the cell surface. These signals are then transmitted inside the cell by second messenger systems that often involved the phosphorylation of proteins.

A very well understood example of extrinsic control is the regulation of glucose metabolism by the hormone insulin. Insulin is produced in response to rises in blood glucose levels. Binding of the hormone to insulin receptors on cells then activates a cascade of protein kinases that cause the cells to take up glucose and convert it into storage molecules such as fatty acids and glycogen. The metabolism of glycogen is controlled by activity of phosphorylase, the enzyme that breaks down glycogen, and glycogen synthase, the enzyme that makes it. These enzymes are regulated in a reciprocal fashion, with phosphorylation inhibiting glycogen synthase, but activating phosphorylase. Insulin causes glycogen synthesis by activating protein phosphatases and producing a decrease in the phosphorylation of these enzymes.

Evolution

The central pathways of metabolism described above, such as glycolysis and the citric acid cycle, are present in all three domains of living things and were present in the last universal ancestor. This universal ancestral cell was prokaryotic and probably a methanogen that had extensive amino acid, nucleotide, carbohydrate and lipid metabolism. The retention of these ancient pathways during later evolution may be the result of these reactions being an optimal solution to their particular metabolic problems, with pathways such as glycolysis and the citric acid cycle producing their end products highly efficiently and in a minimal number of steps. The first pathways of enzyme-based metabolism may have been parts of purine nucleotide metabolism, with previous metabolic pathways being part of the ancient RNA world.

Many models have been proposed to describe the mechanisms by which novel metabolic pathways evolve. These include the sequential addition of novel enzymes to a short ancestral pathway, the duplication and then divergence of entire pathways as well as the recruitment of pre-existing enzymes and their assembly into a novel reaction pathway. The relative importance of these mechanisms is unclear, but genomic studies have shown that enzymes in a pathway are likely to have a shared ancestry, suggesting that many pathways have evolved in a step-by-step fashion with novel functions being created from pre-existing steps in the pathway. An alternative model comes from studies that trace the evolution of proteins' structures in metabolic networks, this has suggested that enzymes are pervasively recruited, borrowing enzymes to perform similar functions in different metabolic pathways (evident in the MANET database) These recruitment processes result in an evolutionary enzymatic mosaic. A third possibility is that some parts of metabolism might exist as "modules" that can be reused in different pathways and perform similar functions on different molecules.

As well as the evolution of new metabolic pathways, evolution can also cause the loss of metabolic functions. For example, in some parasites metabolic processes that are not essential for survival are lost and preformed amino acids, nucleotides and carbohydrates may instead be scavenged from the host. Similar reduced metabolic capabilities are seen in endosymbiotic organisms.

Investigation and Manipulation

Classically, metabolism is studied by a reductionist approach that focuses on a single metabolic pathway. Particularly valuable is the use of radioactive tracers at the whole-organism, tissue and cellular levels, which define the paths from precursors to final products by identifying radioactively labelled intermediates and products. The enzymes that catalyze these chemical reactions can then be purified and their kinetics and responses to inhibitors investigated. A parallel approach is to identify the small molecules in a cell or tissue; the complete set of these molecules is called the metabolome. Overall, these studies give a good view of the structure and function of simple metabolic pathways, but are inadequate when applied to more complex systems such as the metabolism of a complete cell.

An idea of the complexity of the metabolic networks in cells that contain thousands of different enzymes is given by the figure showing the interactions between just 43 proteins and 40 metabolites to the right: the sequences of genomes provide lists containing anything up to 45,000 genes. However, it is now possible to use this genomic data to reconstruct complete networks of biochemical reactions and produce more holistic mathematical models that may explain and predict their behavior. These models are especially powerful when used to integrate the pathway and metabolite data obtained through classical methods with data on gene expression from proteomic and DNA microarray studies. Using these techniques, a model of human metabolism has now been produced, which will guide future drug discovery and biochemical research. These models are now being used in network analysis, to classify human diseases into groups that share common proteins or metabolites.

Bacterial metabolic networks seem to be a striking example of bow-tie organization, an architecture able to input a wide range of nutrients and produce a large variety of products and complex macromolecules using a relatively few intermediate common currencies.

A major technological application of this information is metabolic engineering. Here, organisms such as yeast, plants or bacteria are genetically modified to make them more useful in biotechnology and aid the production of drugs such as antibiotics or industrial chemicals such as 1,3-propanediol and shikimic acid. These genetic modifications usually aim to reduce the amount of energy used to produce the product, increase yields and reduce the production of wastes.

History

The term *metabolism* is derived from the "Metabolismos" for "change", or "overthrow". The history of the scientific study of metabolism spans several centuries and has moved from examining whole animals in early studies, to examining individual metabolic reactions in modern biochemistry. The first controlled experiments in human metabolism were published by Santorio Santorio in 1614 in his book *Ars de statica medicina*. He described how he weighed himself before and after eating, sleep, working, sex, fasting, drinking, and excreting. He found that most of the food he took in was lost through what he called "insensible perspiration".

In these early studies, the mechanisms of these metabolic processes had not been identified and a vital force was thought to animate living tissue. In the 19th century, when studying the fermentation of sugar to alcohol by yeast, Louis Pasteur concluded that fermentation was catalyzed by substances within the yeast cells he called "ferments". He wrote that "alcoholic fermentation is an act correlated with the life and organization of the yeast cells, not with the death or putrefaction of the cells." This discovery, along with the publication by Friedrich Wöhler in 1828 of the chemical synthesis of urea, proved that the organic compounds and chemical reactions found in cells were no different in principle than any other part of chemistry.

It was the discovery of enzymes at the beginning of the 20th century by Eduard Buchner that separated the study of the chemical reactions of metabolism from the biological study of cells, and marked the beginnings of biochemistry. The mass of biochemical knowledge grew rapidly throughout the early 20th century. One of the most prolific of these modern biochemists was Hans Krebs who made huge contributions to the study of metabolism. He discovered the urea cycle and later, working with Hans Kornberg, the citric acid cycle and the glyoxylate cycle. Modern biochemical research has been greatly aided by the development of new techniques such as chromatography, X-ray diffraction, NMR spectroscopy, radioisotopic labelling, electron microscopy and molecular dynamics simulations. These techniques have allowed the discovery and detailed analysis of the many molecules and metabolic pathways in cells.

Between successive cell divisions, cells grow through the functioning of cellular metabolism. Cell metabolism is the process by which individual cells process nutrient molecules. Metabolism has two distinct divisions: catabolism, in which the cell breaks down complex molecules to produce energy and reducing power, and anabolism, in which the cell uses energy and reducing power to construct complex molecules and perform other biological functions. Complex sugars consumed by the organism can be broken down into a less chemically complex sugar molecule called glucose. Once inside the cell, glucose is broken down to make adenosine triphosphate (ATP), a form of energy, through two different pathways.

The first pathway, glycolysis, requires no oxygen and is referred to as anaerobic metabolism. Each reaction is designed to produce some hydrogen ions that can then be used to make energy packets (ATP). In prokaryotes, glycolysis is the only method used for converting energy.

The second pathway, called the Krebs cycle, or citric acid cycle, occurs inside the mitochondria and can generate enough ATP to run all the cell functions.

Within the nucleus of the cell (*light blue*), genes (DNA, *dark blue*) are transcribed into RNA. This RNA is then subject to post-transcriptional modification and control, resulting in a mature mRNA (*red*) that is then transported out of the nucleus and into the cytoplasm (*peach*), where it undergoes translation into a protein. mRNA is translated by ribosomes (*purple*) that match the three-base codons of the mRNA to the three-base anti-codons of the appropriate tRNA. Newly synthesized proteins (*black*) are often further modified, such as by binding to an effector molecule (*orange*), to become fully active.

Creation

Cell division involves a single cell (called a *mother cell*) dividing into two daughter cells. This leads to growth in multicellular organisms (the growth of tissue) and to procreation (vegetative reproduction) in unicellular organisms.

Prokaryotic cells divide by binary fission. Eukaryotic cells usually undergo a process of nuclear division, called mitosis, followed by division of the cell, called cytokinesis. A diploid cell may also undergo meiosis to produce haploid cells, usually four. Haploid cells serve as gametes in multicellular organisms, fusing to form new diploid cells. DNA replication, or the process of duplicating a cell's genome, is required every time a cell divides. Replication, like all cellular activities, requires specialized proteins for carrying out the job.

Protein Synthesis

Cells are capable of synthesizing new proteins, which are essential for the modulation and maintenance of cellular activities. This process involves the formation of new protein molecules from amino acid building blocks based on information encoded in DNA/RNA. Protein synthesis generally consists of two major steps: transcription and translation.

Transcription is the process where genetic information in DNA is used to produce a complementary RNA strand. This RNA strand is then processed to give messenger RNA (mRNA), which is free to migrate through the cell. mRNA molecules bind to protein-RNA complexes called ribosomes located in the cytosol, where they are translated into polypeptide sequences. The ribosome mediates the formation of a polypeptide sequence based on the mRNA sequence. The mRNA sequence directly relates to the polypeptide sequence by binding to transfer RNA (tRNA) adapter molecules in binding pockets within the ribosome. The new polypeptide then folds into a functional three-dimensional protein molecule.

Movement or Motility

Cells can move during many processes: such as wound healing, the immune response and cancer metastasis. For wound healing to occur, white blood cells and cells that ingest bacteria move to the wound site to kill the microorganisms that cause infection.

At the same time fibroblasts (connective tissue cells) move there to remodel damaged structures. In the case of tumour development, cells from a primary tumour move away and spread to other parts of the body. Cell motility involves many receptors, crosslinking, bundling, binding, adhesion, motor and other proteins. The process is divided into three steps – protrusion of the leading edge of the cell, adhesion of the leading edge and de-adhesion at the cell body and rear, and cytoskeletal contraction to pull the cell forward. Each step is driven by physical forces generated by unique segments of the cytoskeleton.

Evolution

The origin of cells has to do with the origin of life, which began the history of life on Earth.

Origin of the First Cell

There are several theories about the origin of small molecules that could lead to life in an early Earth. One is that they came from meteorites. Another is that they were created at deep-sea vents. A third is that they were synthesized by lightning in a reducing atmosphere; although it is not clear if Earth had such an atmosphere. There are essentially no experimental data defining what the first self-replicating forms were. RNA is generally assumed to be the earliest self-replicating molecule, as it is capable of both storing genetic information and catalyzing chemical reactions. But some other entity with the potential to self-replicate could have preceded RNA, like clay or peptide nucleic acid.

Cells emerged at least 4.0–4.3 billion years ago. The current belief is that these cells were heterotrophs. An important characteristic of cells is the cell membrane, composed of a bilayer of lipids. The early cell membranes were probably more simple and permeable than modern ones, with only a single fatty acid chain per lipid. Lipids are known to spontaneously form bilayered vesicles in water, and could have preceded RNA. But the first cell membranes could also have been produced by catalytic RNA, or even have required structural proteins before they could form.

Origin of Eukaryotic Cells

The eukaryotic cell seems to have evolved from a symbiotic community of prokaryotic cells. DNA-bearing organelles like the mitochondria and the chloroplasts are almost certainly what remains of ancient symbiotic oxygen-breathing proteobacteria and cyanobacteria, respectively, where the rest of the cell seems to be derived from an ancestral archaean prokaryote cell – a theory termed the endosymbiotic theory.

There is still considerable debate about whether organelles like the hydrogenosome predated the origin of mitochondria, or viceversa: see the hydrogen hypothesis for the origin of eukaryotic cells.

Sex, as the stereotyped choreography of meiosis and syngamy that persists in nearly all extant eukaryotes, may have played a role in the transition from prokaryotes to eukaryotes. An 'origin of sex as vaccination' theory suggests that the eukaryote genome accreted from prokaryan parasite genomes in numerous rounds of lateral gene transfer. Sex-as-syngamy (fusion sex) arose when infected hosts began swapping nuclearized genomes containing co-evolved, vertically transmitted symbionts that conveyed protection against horizontal infection by more virulent symbionts.

4

Genetic Engineering or Recombinant DNA Technology

Genetic engineering, also called genetic modification, is the direct human manipulation of an organism's genetic material in a way that does not occur under natural conditions. It involves the use of recombinant DNA techniques, but does not include traditional animal and plant breeding or mutagenesis. Any organism that is generated using these techniques is considered to be a genetically modified organism. The first organisms genetically engineered were bacteria in 1973 and then mice in 1974. Insulin producing bacteria were commercialized in 1982 and genetically modified food has been sold since 1994. The most common form of genetic engineering involves the insertion of new genetic material at an unspecified location in the host genome. This is accomplished by isolating and copying the genetic material of interest, generating a construct containing all the genetic elements for correct expression, and then inserting this construct into the host organism. Other forms of genetic engineering include gene targeting and knocking out specific genes via engineered nucleases such as zinc finger nucleases or engineered homing endonucleases.

Genetic engineering techniques have been applied in numerous fields including research, biotechnology, and medicine. Medicines such as insulin and human growth hormone are now produced in bacteria, experimental mice such as the oncomouse and the knockout mouse are being used for research purposes and insect resistant and/or herbicide tolerant crops have been commercialized. Genetically engineered plants and animals capable of producing biotechnology drugs more cheaply than current methods (called pharming) are also being developed and in 2009 the FDA approved the sale of the pharmaceutical protein antithrombin produced in the milk of genetically engineered goats.

Definition

Genetic engineering alters the genetic makeup of an organism using techniques that introduce heritable material prepared outside the organism either directly into the host or into a cell that is then fused or hybridized with the host. This involves using recombinant nucleic acid (DNA or RNA) techniques to form new combinations of heritable genetic material followed by the incorporation of that material either indirectly through a vector system or directly through micro-injection, macro-injection and micro-encapsulation techniques. Genetic engineering does not include traditional animal and plant breeding, in vitro fertilisation, induction of polyploidy, mutagenesis and cell fusion techniques that do not use recombinant nucleic acids or a genetically modified organism in the process. Cloning and stem cell research, although not considered genetic engineering, are closely related and genetic engineering can be used within them. Synthetic biology is an emerging discipline that takes genetic engineering a step further by introducing artificially synthesized genetic material from raw materials into an organism.

If genetic material from another species is added to the host, the resulting organism is called transgenic. If genetic material from the same species or a species that can naturally breed with the host is used the resulting organism is called cisgenic. Genetic engineering can also be used to remove genetic material from the target organism, creating a knock out organism. In Europe genetic modification is synonymous with genetic engineering while within the United States of America it can also refer to conventional breeding methods.

History

Humans have altered the genomes of species for thousands of years through artificial selection and more recently mutagenesis. Genetic engineering as the direct manipulation of DNA by humans outside breeding and mutations has only existed since the 1970s. The term "genetic engineering" was first coined by Jack Williamson in his science fiction novel *Dragon's Island*, published in 1951, one year before DNA's role in heredity was confirmed by Alfred Hershey and Martha Chase, and two years before James Watson and Francis Crick showed that the DNA molecule has a double-helix structure.

In 1972 Paul Berg created the first recombinant DNA molecules by combined DNA from the monkey virus SV40 with that of the lambda virus. In 1973 Herbert Boyer and Stanley Cohen created the first transgenic organism by inserting antibiotic resistance genes into the plasmid of an *E. coli* bacterium. A year later Rudolf Jaenisch created a transgenic mouse by introducing foreign DNA into its embryo, making it the world's first transgenic animal. In 1976 Genentech, the first genetic engineering company was founded by Herbert Boyer and Robert Swanson and a year later and the company produced a human protein (somatostatin) in *E.coli*. Genentech announced the production of genetically engineered human insulin in 1978.

In 1980, the U.S. Supreme Court in the Diamond v. Chakrabarty case ruled that genetically altered life could be patented. The insulin produced by bacteria, branded humulin, was approved for release by the Food and Drug Administration in 1982. The first field trials of genetically engineered plants occurred in France and the USA in 1986, tobacco plants were engineered to be resistant to herbicides. The People's Republic of China was the first country to commercialize transgenic plants, introducing a virus-resistant tobacco in 1992.

In 1994 Calgene attained approval to commercially release the Flavr Savr tomato, a tomato engineered to have a longer shelf life. In 1994, the European Union approved tobacco engineered to be resistant to the herbicide bromoxynil, making it the first genetically engineered crop commercialized in Europe. In 1995, Bt Potato was approved safe by the Environmental Protection Agency, making it the first pesticide producing crop to be approved in the USA. In 2009 11 transgenic crops were grown commercially in 25 countries, the largest of which by area grown were the USA, Brazil, Argentina, India, Canada, China, Paraguay and South Africa.

In 2010, scientists at the J. Craig Venter Institute, announced that they had created the first synthetic bacterial genome, and added it to a cell containing no DNA. The resulting bacterium, named Synthia, was the world's first synthetic life form.

Isolating the Gene

First, the gene to be inserted into the genetically modified organism must be chosen and isolated. Presently, most genes transferred into plants provide protection against insects or tolerance to herbicides. In animals the majority of genes used are growth hormone genes.

Once chosen the genes must be isolated. This typically involves multiplying the gene using polymerase chain reaction (PCR). If the chosen gene or the donor organism's genome has been well studied it may be present in a genetic library. If the DNA sequence is known, but no copies of the gene are available, it can be artificially synthesized. Once isolated, the gene is inserted into a bacterial plasmid.

Constructs

The gene to be inserted into the genetically modified organism must be combined with other genetic elements in order for it to work properly. The gene can also be modified at this stage for better expression or effectiveness. As well as the gene to be inserted most constructs contain a promoter and terminator region as well as a selectable marker gene. The promoter region initiates transcription of the gene and can be used to control the location and level of gene expression, while the terminator region ends transcription. The selectable marker, which in most cases confers antibiotic resistance to the organism it is expressed in, is needed to determine which cells are transformed with the new gene. The constructs are made using recombinant DNA techniques, such as restriction digests, ligations and molecular cloning.

Gene Targeting

The most common form of genetic engineering involves inserting new genetic material randomly within the host genome. Other techniques allow new genetic material to be inserted at a specific location in the host genome or generate mutations at desired genomic loci capable of knocking out endogenous genes. The technique of gene targeting uses homologous recombination to target desired changes to a specific endogenous gene. This tends to occur at a relatively low frequency in plants and animals and generally requires the use of selectable markers. The frequency of gene targeting can be greatly enhanced with the use of engineered nucleases such as zinc finger nucleases, engineered homing endonucleases, or nucleases created from TAL effectors. In addition to enhancing gene targeting, engineered nucleases can also be used to introduce mutations at endogenous genes that generate a gene knockout.

Transformation

About 1% of bacteria are naturally able to take up foreign DNA but it can also be induced in other bacteria. Stressing the bacteria for example, with a heat shock or an electric shock, can make the cell membrane permeable to DNA that may then incorporate into their genome or exist as extrachromosomal DNA. DNA is generally inserted into animal cells using microinjection, where it can be injected through the cells nuclear envelope directly into the nucleus or through the use of viral vectors. In plants the DNA is generally inserted using *Agrobacterium*-mediated recombination or biolistics.

In *Agrobacterium*-mediated recombination the plasmid construct must also contain T-DNA. *Agrobacterium* naturally inserts DNA from a tumour inducing plasmid into any susceptible plant's genome it infects, causing crown gall disease. The T-DNA region of this plasmid is responsible for insertion of the DNA. The genes to be inserted are cloned into a binary vector, which contains T-DNA and can be grown in both *E. Coli* and *Agrobacterium*. Once the binary vector is constructed the plasmid is transformed into *Agrobacterium* containing no plasmids and plant cells are infected. The *Agrobacterium* will then naturally insert the genetic material into the plant cells.

In biolistics particles of gold or tungsten are coated with DNA and then shot into young plant cells or plant embryos. Some genetic material will enter the cells and transform them. This method can be used on plants that are not susceptible to *Agrobacterium* infection and also allows transformation of plant plastids. Another transformation method for plant and animal cells is electroporation. Electroporation involves subjecting the plant or animal cell to an electric shock, which can make the cell membrane permeable to plasmid DNA. In some cases the electroporated cells will incorporate the DNA into their genome. Due to the damage caused to the cells and DNA the transformation efficiency of biolistics and electroporation is lower than agrobacterial mediated transformation and microinjection.

Selection

Not all the organism's cells will be transformed with the new genetic material; in most cases a selectable marker is used to differentiate transformed from untransformed cells. If a cell has been successfully transformed with the DNA it will also contain the marker gene. By growing the cells in the presence of an antibiotic or chemical that selects or marks the cells expressing that gene it is possible to separate the transgenic events from the non-transgenic. Another method of screening involves using a DNA probe that will only stick to the inserted gene. A number of strategies have been developed that can remove the selectable marker from the mature transgenic plant.

Regeneration

As often only a single cell is transformed with genetic material the organism must be regrown from that single cell. As bacteria consist of a single cell and reproduce clonally regeneration is not necessary. In plants this is accomplished through the use of tissue culture. Each plant species has different requirements for successful regeneration through tissue culture. If successful an adult plant is produced that contains the transgene in every cell. In animals it is necessary to ensure that the inserted DNA is present in the embryonic stem cells. When the offspring is produced they can be screened for the presence of the gene. All offspring from the first generation will be heterozygous for the inserted gene and must be mated together to produce a homozygous animal.

Confirmation

Further tests using PCR, Southern Blots and Bioassays are needed to confirm that the gene is expressed and functions correctly. The organism's offspring are also tested to ensure that the trait can be inherited and that it follows a Mendelian inheritance pattern.

Applications

Genetic engineering has applications in medicine, research, industry and agriculture and can be used on a wide range of plants, animals and micro organism.

Medicine

In medicine genetic engineering has been used to mass-produce insulin, human growth hormones, follistim (for treating infertility), human albumin, monoclonal antibodies, antihemophilic factors, vaccines and many other drugs. Vaccination generally involves injecting weak live, killed or inactivated forms of viruses or their toxins into the person being immunized. Genetically engineered viruses are being developed that can still confer immunity, but lack the infectious sequences. Mouse hybridomas, cells fused together to create monoclonal antibodies, have been humanised through genetic engineering to create human monoclonal antibodies.

Genetic engineering is used to create animal models of human diseases. Genetically modified mice are the most common genetically engineered animal model. They have been used to study and model cancer (the oncomouse), obesity, heart disease, diabetes, arthritis, substance abuse, anxiety, aging and Parkinson disease. Potential cures can be tested against these mouse models. Also genetically modified pigs have been bred with the aim of increasing the success of pig to human organ transplantation. Gene therapy is the genetic engineering of humans by replacing defective human genes with functional copies. This can occur in somatic tissue or germline tissue. If the gene is inserted into the germline tissue it can be passed down to that person's descendants. Gene therapy has been used to treat patients suffering from immune deficiencies (notably Severe combined immunodeficiency) and trials have been carried out on other genetic disorders. The success of gene therapy so far has been limited and a patient (Jesse Gelsinger) has died during a clinical trial testing a new treatment. There are also ethical concerns should the technology be used not just for treatment, but for enhancement, modification or alteration of a human beings' appearance, adaptability, intelligence, character or behavior. The distinction between cure and enhancement can also be difficult to establish. Transhumanists consider the enhancement of humans desirable.

Research

Genetic engineering is an important tool for natural scientists. Genes and other genetic information from a wide range of organisms are transformed into bacteria for storage and modification, creating genetically modified bacteria in the process. Bacteria are cheap, easy to grow, clonal, multiply quickly, relatively easy to transform and can be stored at -80°C almost indefinitely. Once a gene is isolated it can be stored inside the bacteria providing an unlimited supply for research.

Organisms are genetically engineered to discover the functions of certain genes. This could be the effect on the phenotype of the organism, where the gene is expressed or what other genes it interacts with. These experiments generally involve loss of function, gain of function, tracking and expression.

- Loss of function experiments, such as in a gene knockout experiment, in which an organism is engineered to lack the activity of one or more genes. A knockout experiment involves the creation and manipulation of a DNA construct *in vitro*, which, in a simple knockout, consists of a copy of the desired gene, which has been altered such that it is non-functional. Embryonic stem cells incorporate the altered gene, which replaces the already present functional copy. These stem cells are injected into blastocysts, which are implanted into surrogate mothers. This allows the experimenter to analyze the defects caused by this mutation and thereby determine the role of particular genes. It is used especially frequently in developmental biology. Another method, useful in organisms such as Drosophila (fruit fly), is to induce mutations in a large population and then screen the progeny for the desired mutation. A similar process can be used in both plants and prokaryotes.

- Gain of function experiments, the logical counterpart of knockouts. These are sometimes performed in conjunction with knockout experiments to more finely establish the function of the desired gene. The process is much the same as that in knockout engineering, except that the construct is designed to increase the function of the gene, usually by providing extra copies of the gene or inducing synthesis of the protein more frequently.

- Tracking experiments, which seek to gain information about the localization and interaction of the desired protein. One way to do this is to replace the wild-type gene with a 'fusion' gene, which is a juxtaposition of the wild-type gene with a reporting element such as green fluorescent protein (GFP) that will allow easy visualization of the products of the genetic modification. While this is a useful technique, the manipulation can destroy the function of the gene, creating secondary effects and possibly calling into question the results of the experiment. More sophisticated techniques are now in development that can track protein products without mitigating their function, such as the addition of small sequences that will serve as binding motifs to monoclonal antibodies.

- Expression studies aim to discover where and when specific proteins are produced. In these experiments, the DNA sequence before the DNA that codes for a protein, known as a gene's promoter, is reintroduced into an organism with the protein coding region replaced by a reporter gene such as GFP or an enzyme that catalyzes the production of a dye. Thus the time and place where a particular protein is produced can be observed. Expression studies can be taken a step further by altering the promoter to find which pieces are crucial for the proper expression of the gene and are actually bound by transcription factor proteins; this process is known as promoter bashing.

Industrial

By engineering genes into bacterial plasmids it is possible to create a biological factory that can produce proteins and enzymes. Some genes do not work well in bacteria, so yeast, a eukaryote, can also be used. Bacteria and yeast factories have been used to produce medicines such as insulin, human growth hormone, and vaccines, supplements such as tryptophan, aid in the production of food (chymosin in cheese making) and fuels. Other applications involving genetically engineered bacteria being investigated involve making the bacteria perform tasks outside their natural cycle, such as cleaning up oil spills, carbon and other toxic waste.

Agriculture

One of the best-known and controversial applications of genetic engineering is the creation of genetically modified food. There are three generations of genetically modified crops. First generation crops have been commercialized and most provide protection from insects and/or resistance to herbicides. There are also fungal and virus resistant crops developed or in development. They have been developed to make the insect and weed management of crops easier and can indirectly increase crop yield.

The second generation of genetically modified crops being developed aim to directly improve yield by improving salt, cold or drought tolerance and to increase the nutritional value of the crops. The third generation consists of pharmaceutical crops, crops that contain edible vaccines and other drugs. Some agriculturally important animals have been genetically modified with growth hormones to increase their size while others have been engineered to express drugs and other proteins in their milk.

The genetic engineering of agricultural crops can increase the growth rates and resistance to different diseases caused by pathogens and parasites. This is beneficial as it can greatly increase the production of food sources with the usage of fewer resources that would be required to host the world's growing populations. These modified crops would also reduce the usage of chemicals, such as fertilizers and pesticides, and therefore decrease the severity and frequency of the damages produced by these chemical pollution. Ethical and safety concerns have been raised around the use of genetically modified food. A major safety concern relates to the human health implications of eating genetically modified food, in particular whether toxic or allergic reactions could occur. Gene flow into related non-transgenic crops, off target effects on beneficial organisms and the impact on biodiversity are important environmental issues. Ethical concerns involve religious issues, corporate control of the food supply, intellectual property rights and the level of labelling needed on genetically modified products.

Other Uses

In materials science, a genetically modified virus has been used to construct a more environmentally friendly lithium-ion battery. Some bacteria have been genetically engineered to create black and white photographs while others have potential to be used as sensors by expressing a fluorescent protein under certain environmental conditions. Genetic engineering is also being used to create BioArt and novelty items such as blue roses, and glowing fish.

Opposition and Criticism

A 2010 study of Canola found transgenes in 80% of wild (uncultivated or "feral") varieties in North Dakota, meaning 80% of the plants which had established themselves in the area were genetically engineered varieties. The researchers stated that "we found the highest densities of [such transgene-containing] plants near agricultural fields and along major freeways, but we were also finding plants in the middle of nowhere" adding that "over time,...the build-up of different types of herbicide resistance in feral [natural] canola and closely related weeds, like field mustard, could make it more difficult to manage these plants using herbicides."

Human Genetic Engineering

Human genetic engineering is the alteration of an individual's genotype with the aim of choosing the phenotype of a newborn or changing the existing phenotype of a child or adult. It holds the promise of curing genetic diseases like cystic fibrosis, and increasing the immunity of people to viruses. It is speculated that genetic engineering could be used to change physical appearance, metabolism, and even improve mental faculties like memory and intelligence, although for now these uses seem to be of lower priority to researchers and are therefore limited to science fiction.

History

The first gene therapy trials on humans began in 1990 on patients with Severe Combined Immunodeficiency (SCID). In 2000, the first gene therapy "success" resulted in SCID patients with a functional immune system. These trials were stopped when it was discovered that two of ten patients in one trial had developed leukemia resulting from the insertion of the gene-carrying retrovirus near an oncogene. In 2007, four of the ten patients had developed leukemia. Work is now focusing on correcting the gene without triggering an oncogene.

Trial treatments of SCID have been gene therapy's only success; since 1999, gene therapy has restored the immune systems of at least 17 children with two forms (ADA-SCID and X-SCID) of the disorder.

Human genetic engineering is already being used on a small scale to allow infertile women with genetic defects in their mitochondria to have children. Healthy human eggs from a second mother are used. The child produced this way has genetic information from two mothers and one father. The changes made are germline changes and will likely be passed down from generation to generation, and, thus, are a permanent change to the human genome.

Other forms of human genetic engineering are still theoretical. Recombinant DNA research is usually performed to study gene expression and various human diseases. Some drastic demonstrations of gene modification have been made with mice and other animals, however, testing on humans is generally considered off-limits. In some instances changes are usually brought about by removing genetic material from one organism and transferring them into another species.

Methods

Somatic: Somatic genetic engineering involves adding genes to cells other than egg or sperm cells. For example, if a person had a disease caused by a defective gene, a healthy gene could be added to the affected cells to treat the disorder. As of now, this is likely to take the form of gene therapy. The distinguishing characteristic of somatic engineering is that it is non-inheritable, i.e. the new gene would not be passed to the recipient's offspring.

There are two techniques researchers are currently experimenting with:

- Viruses are good at injecting their DNA payload into human cells and reproducing it. By adding the desired DNA to the DNA of non-pathogenic virus, a small amount of virus will reproduce the desired DNA and spread it all over the body.

- Manufacture large quantities of DNA, and somehow package it to induce the target cells to accept it, either as an addition to one of the original 23 chromosomes, or as an independent 24th human artificial chromosome.

Germline

Germline engineering involves changing genes in eggs, sperm, or very early embryos. This type of engineering is inheritable, meaning that the modified genes would appear not only in offspring that resulted from the procedure, but also in subsequent generations.

Uses

Two motivators of human genetic engineering are referred to as "negative" and "positive". The former aims to remove genetic disorders and the latter aims to alter phenotypic expression to result in an enhanced being.

Negative Genetic Engineering (Cures and Treatments)

When treating problems that arise from genetic disorder, one solution is gene therapy, also known as negative genetic engineering. A genetic disorder is a condition caused by the genetic code of the individual, such as spina bifida or autism. When this happens, genes may be expressed in unfavorable ways or not at all, and this generally leads to further complications.

The idea of gene therapy is that a non-pathogenic virus or other delivery systems can be used to insert into DNA—a good copy of the gene—into cells of the living individual. The modified cells would divide as normal and each division would produce cells that express the desired trait. The result would be that he/she would then have the ability to express the trait that was previously absent, at least partially. This form of genetic engineering could help alleviate many problems, such as diabetes, cystic fibrosis, or other genetic diseases.

Positive Genetic Engineering (Enhancement)

The potential of genetic engineering to cure medical conditions opens the question of exactly what such a condition is. Some view aging and death as medical conditions and therefore potential targets for engineering solutions. They see human genetic engineering potentially as a key tool in this. The difference between cure and enhancement from this perspective is merely one of degree. Theoretically genetic engineering could be used to drastically change people's genomes, which could enable people to regrow limbs and other organs, perhaps even extremely complex ones such as the spine.

It could also be used to make people smarter, stronger, or to increase the capacity of the lungs, among other things. If a gene exists in nature, perhaps it could be changed into a human cell. In this view, there is no qualitative difference (only a quantitative one) between, for instance, a genetic intervention to cure muscular atrophy, and a genetic intervention to improve muscle function even when those muscles are functioning at or below the human average (since there is also an average muscle function for those with a particular type of dystrophy, which the treatment would improve upon).

Others feel, there is an important distinction between using genetic technologies to treat those who are suffering, and to make those who are already healthy seem more superior to the average person. Though theory and speculation suggest, that genetic engineering could be used to make people stronger, faster, smarter, or to increase lung capacity. The AAAS report finds that there is little evidence to support this theory. Can this currently be done without very unsafe and therefore unethical human experiments. Because different cells have different tasks, changing one cell to do a function differently, will not only affect that one task, but it can affect many other tasks as well.

Controversy

The genetic engineering of humans has raised many controversial ethical issues. While negative genetic engineering (gene therapy) does indeed raise a debate, the use of genetic engineering for human enhancement arouses the strongest feelings on both sides.

Genetic engineering is tested on animals, often including primates. Some animal rights activists find this inhumane.

Genetic engineering can be used to cure peoples with diabetes. It is possible to extract genes from cells which are called beta cells and then to insert the insulin producing genes into a bacterium. Then the bacterium will start producing insulin. Genetic modification of embryos can pose an ethical question about the rights of the baby. One belief is that every fetus should be free to not be genetically modified. Others believe that parents hold the rights to change their unborn children. Still others believe that every child should have the right to be born free from preventable diseases.

Molecular Biologist Lee M. Silver believes that unlike Aldous Huxley's Brave New World, where a totalitarian government controls all of the genetic enhancements (they actually use eugenics instead of direct genetic modification) in society, the use of gene therapy to design children will be spread through what he calls "free market eugenics" (Silver 315). Wealthy families will opt to design their child with genetic advantages because other families are doing so, and everybody wants to provide their newborn child with the best opportunities in life, with a leg up on the competition.

The greatest fear for Silver is that we will design so many children with germline gene therapy, that the families wealthy enough to design their children, will pass down these enhanced traits to future generations. This gene therapy will obviously cost money, and the less wealthy families will be left to procreate naturally, and introduce their children into the world disadvantaged from their first breath.

The impact on society will be a new alignment of classes, no longer will we separate people by their ethnic differences, the new division will be between what Silver calls 'the naturals' and 'the GenRich', or genetically enhanced. The major worry here is that the 'genetic gulf' between these two classes will become so wide that humans will become separate species (Silver 313).

In Popular Culture

- *Maximum Ride series by James Patterson:* The main characters are six human children who had bird DNA injected into them while they were in their mothers' wombs.

- *Mobile Suit Gundam SEED (anime):* Set in a world in which genetically modified humans, termed 'Coordinators', have been ostracized and isolated from unmodified humans, termed 'Naturals'. Due to extreme differences in mental and physical abilities between the two groups, racial, economic, and political issues have arisen, culminating in war. Gundam Seed addresses such concerns as animosity caused by the jealousy of Naturals over Coordinator abilities, both groups looking down on one another as being lesser life forms, and genocidal factions emerging on both sides. The series primarily explores these issues from the point of view of a Coordinator protagonist who finds himself fighting on the side of the Naturals, as his childhood friend has become a member of the Coordinator military, giving a perspective on both sides of the conflict.

- *Gattaca* (film): Presents a biopunk vision of a society driven by new eugenics. Children of the middle and upper classes are selected through preimplantation genetic diagnosis to ensure they possess the best hereditary traits of their parents.

- *Oryx and Crake by Margaret Atwood:* Apocalyptic pseudo-dystopian story, one of the chief plot points of which involves the genetic engineering of a new type of transhuman and the pathological decimation of *Homo sapiens.*

- *Bioshock (Video Game):* The main enemies the player encounters over the course of the game are known as splicers, so called for their genetic manipulation, or gene splicing. In the game a compound called ADAM is responsible for the genetic manipulation. The compound is harvested from a species of sea slugs, it acts like a seemingly benign form of cancer, destroying native cells and replacing them with the unstable stem versions. The splicers have all become so addicted to the ADAM that they will kill anything or anyone for it.

- *Old Man's War series by John Scalzi:* To create an army of soldiers capable of defending the human race from endless alien hordes the Colonial Defense Forces recruits 75 year olds and gives them new younger bodies capable of super human feats.

- *Batman Beyond:* In the episode *Splicers* and making appearances later in the series, a new fad of splicing animal genes for cosmetic and enhancement purposes makes the teen scene. When research shows that splicing increases aggression in users, resulting it being banned in Gotham, only finding a place in the criminal underworld later. Notable Splicers are Woof of the Jokerz (spliced with hyena DNA), and the Zander (spliced with T-Rex DNA), leader of KOBRA.

- *Halo (series) (Video Game):* The protagonist and several other characters in the Halo universe known as Spartans have all gone through comprehensive genetic augmentation, making them almost super human. The augmentations vary from Muscular Enhancement Injections to a Catalytic Thyroid Implant. The effects include nearly unbreakable bones, increased reflexes and increased muscle tissue with increased density. The SPARTAN-III program also included a mutagen that increased aggression and injury tolerance.

- *Deus Ex:* The protagonists of both games in the series are genetically modified by injections called nano-augs (Deus Ex) and bio mods (Invisible War). These nanite injections alter the host's genes to enhance them with new skill that would help them through different obstacles in the game. Both the protagonists and antagonists are bio-modified along with other support characters and neutral characters.

- *PROTOTYPE:* The story is centered around a genetic engineering company called Gentek. The engineers in Gentek modified a chimeran-virus to make it 10 times as deadly, to 'unlock' previously dormant parts of a subjects DNA, and they made it to the point of where it could copy those infected with it down to the genetic level. The new 'infected' being then has complete and total control over its genetic structure, allowing it to develop instantaneous shape-shifting abilities. However, the game reveals that the only 'infected' being that can do this is a person already dead (or dying) when the virus enters the bloodstream. The virus must also be in a state of emergency to survive and not die out.

- *Warhammer 40,000:* In the games, and table-top series, Space marines are the result of Genetic Engineering, by replacing organs, and limbs that have been engineered. The Marine himself, also goes through rigorous physical and mental training, to make him accustom to the newly added appendages and organs.

GM Food Controversy

The genetically modified foods controversy is a dispute over the relative advantages and disadvantages of genetically modified (GM) food crops and other uses of genetically-modified organisms in food production. The dispute involves biotechnology companies, governmental regulators, non-governmental organizations and scientists. The dispute is most intense in Japan and Europe, where public concern about GM food is higher than in other parts of the world such as the United States. In the United States GM crops are more widely grown and the introduction of these products has been less controversial. The five key areas of political controversy related to genetically engineered food are food safety, the effect on natural ecosystems, gene flow into non GE crops, moral/religious concerns, and corporate control of the food supply. To date, not a single instance of harm to human health or the environment has been documented with GM crops. Several benefits have been widely accepted and are uncontested in the scientific literature. These include reductions in insecticide use on GE cotton, enhanced biological diversity in GE cotton fields (compared to non-GE fields), enhanced farmer income and communal benefits, increased yields for poor farmers and improved health of farmworkers. Although the use of herbicide tolerant crops remain controversial, because of the need to spray herbicides, it is clear that the use of these crops has promoted a shift to less toxic herbicides.

The concept of substantial equivalence has been established to address the fears of some consumers about the safety of GE crops. "Substantial equivalence embodies the concept that if a new food or food component is found to be substantially equivalent to an existing food or food component, it can be treated in the same manner with respect to safety (i.e., the food or food component can be concluded to be as safe as the conventional food or food component)" (Joint FAO/ WHO Biotechnology and Food Safety Report, 1996).

The rationale for this approach is that it would be impossible to test all the new crop varieties every year for food safety. Only a few food crops on the market have been shown to cause adverse health effects and all of these were the result of conventional genetic modification, not from genetic engineering To date no adverse health effects caused by products approved for sale have been documented, although two products failed initial safety testing and were discontinued, due to allergic reactions. Most feeding trials have observed no toxic effects and saw that GM foods were equivalent in nutrition to unmodified foods, although a few non-peer-reviewed reports speculate physiological changes to GM food. Although there is now broad scientific consensus that GE crops on the market are safe to eat, some scientists and advocacy groups such as Greenpeace and World Wildlife Fund call for additional and more rigorous testing before marketing genetically engineered food.

Another area of controversy is what effect pest and herbicide-resistant crops have on ecosystems. For example, whereas 14 years of studies indicate cotton engineered for resistance to insects has resulted in massive reduction in insecticides globally, the use of crops engineered for herbicide tolerance has changed herbicide patterns, with a clear shift to use of less toxic herbicides. Studies assessing diversity of non-target insects has revealed enhanced biological diversity in GE fields compared to non-GE fields. The risk and effects of have also been cited as concerns, with the possibility that genes might spread from modified crops to wild relatives. There is no evidence that the risk of pollen transfer is greater for any of the GE crops on the market as compared to non-GE crops.

Health Risks and Benefits

Present Knowledge on GM Food Safety: Worldwide, there is a range of perspectives within non-governmental organizations on the safety of GM foods. For example, the US pro-GM group AgBioWorld has argued that GM foods have been proven safe, while other pressure groups and consumer rights groups, such as the Organic Consumers Association, and Greenpeace claim the long-term health risks which GM could pose, or the environmental risks associated with GM, have not yet been adequately investigated. In Japan, Consumers Union of Japan are opposed to GMO foods. They also claim that truly independent research in these areas is systematically blocked by the GM corporations which own the GM seeds and reference materials.

A 2008 review published by the Royal Society of Medicine noted that GM foods have been eaten by millions of people worldwide for over 15 years, with no reports of ill effects. Similarly a 2004 report from the US National Academies of Sciences stated: "To date, no adverse health effects attributed to genetic engineering have been documented in the human population." A 2004 review of feeding trials in the *Italian Journal of Animal Science* found no differences among animals eating genetically modified plants.

A 2005 review in *Archives of Animal Nutrition* concluded that first-generation genetically modified foods had been found to be similar in nutrition and safety to non-GM foods, but noted that second-generation foods with "significant changes in constituents" would be more difficult to test, and would require further animal studies. However, a 2009 review in *Nutrition Reviews* found that although most studies concluded that GM foods do not differ in nutrition or cause any detectable toxic effects in animals, some studies did report adverse changes at a cellular level caused by some GM foods, concluding that "More scientific effort and investigation is needed to ensure that consumption of GM foods is not likely to provoke any form of health problem". A review published in 2009 by Dona and Arvanitoyannis concluded that "results of most studies with GM foods indicate that they may cause some common toxic effects such as hepatic, pancreatic, renal, or reproductive effects and may alter the hematological, biochemical, and immunologic parameters".

However responses to this review in 2009 and 2010 note that the Dona and Arvanitoyannis concentrated on articles with an anti-GM bias that have been refuted by scientists in peer-reviewed articles elsewhere - for example the 35S promoter, stability of transgenes, antibiotic marker genes and the claims for toxic effects of GM foods.

In 2007, a review by Domingo of the toxicity by searching in the Publimed data base using 12 search terms, cited 68 references, found that the "number of references" on the safety of GM/transgenic crops was "surprisingly limited" and questioned whether the safety of genetically modified food has been demonstrated; the review also remarked that its conclusions were in agreement with three earlier reviews by Zdunczyk (2001), Bakshi (2003), and Pryme and Lembcke (2003). However, an article in 2007 by Vain found 692 research studies focusing on GM crop and food safety and identified a strong increase in the publication of such articles in recent years. Vain commented that the multidisciplinarian nature of GM research complicates the retrieval of GM studies and requires using many search terms (he used more than 300) and multiple data bases.

Safety Assessments

The starting point for the safety assessment of genetically engineered food products is to assess if the food is "substantially equivalent" to its natural counterpart. To decide if a modified product is substantially equivalent, the product is tested by the manufacturer for unexpected changes in a limited set of components such as toxins, nutrients or allergens that are present in the unmodified food. If these tests show no significant difference between the modified and unmodified products, then no further food safety testing is required. The manufacturers data is then assessed by an independent regulatory body, such as the U.S. Food and Drug Administration.

However, if the product has no natural equivalent, or shows significant differences from the unmodified food, then further safety testing is carried out. A 2003 review in *Trends in Biotechnology* identified 7 main parts of a standard safety test:

1. Study of the introduced DNA and the new proteins or metabolites that it produces;

2. Analysis of the chemical composition of the relevant plant parts, measuring nutrients, anti-nutrients as well as any natural toxins or known allergens;

3. Assess the risk of gene transfer from the food to microorganisms in the human gut;

4. Study the possibility that any new components in the food might be allergens;

5. Estimate how much of a normal diet the food will make up;

6. Estimate any toxicological or nutritional problems revealed by this data;

7. Additional animal toxicity tests if there is the possibility that the food might pose a risk.

This process was examined further in a review published by Kuiper *et al.* 2002 in the journal *Toxicology*, which stated that substantial equivalence does not itself measure risks, but instead identifies differences between existing products and new foods, which might pose dangers to health. If differences do exist, identifying these differences is a starting point for a full safety assessment, rather than an end point. The authors concluded that "The concept of substantial equivalence is an adequate tool in order to identify safety issues related to genetically modified products that have a traditional counterpart".

However, the review also noted difficulties in applying this standard in practice, including the fact that traditional foods contain many chemicals that have toxic or carcinogenic effects and that our existing diets therefore have not been proven to be safe. This lack of knowledge on unmodified food poses a problem, as GM foods may have differences in anti-nutrients and natural toxins that have never been identified in the original plant, raising the possibility that harmful changes could be missed.

The application of substantial equivalence has also been more strongly criticized. For example, in a speech in 1999, Andrew Chesson of the University of Aberdeen, stated that substantial equivalence testing "could be flawed in some cases" and that some current safety tests could allow harmful substances to enter the human food chain. In a commentary in *Nature* Millstone *et al.* argued that all GM foods should have extensive biological, toxicological and immunological tests and that the concept of substantial equivalence based solely on chemical analyzes of the components of a food should be abandoned. They stated that this is necessary since it is currently impossible to predict the biological properties of a substance only from knowledge of its chemistry. This commentary was controversial and was criticized for misleading presentation of data and presenting an over-simplified version of safety assessments. For example, Kuiper *et al.* responded to this criticism by noting that equivalency testing does involve more than chemical tests and may include toxicity testing.

Medical writer Barbara Keeler and Marc Lappé argued in a 2001 article in the Los Angeles Times that the differences between Genetically Modified and conventional foods challenges the presumption of equivalence. Using Roundup ready soy that has been on the market since 1995 as an example, they noted the differences when compared to its unmodified counterpart. Significantly lower levels of protein than unmodified soy.

Significantly lower levels of phenylalanine, an essential amino acid and as a dietary supplement, the reason doctors advise the consumption of soy products. Levels of trypsin inhibitor were 27% higher and after toasting, Lectin was double that found in conventional soy, both are known allergens. GM soy also has 29% less choline, a B-complex vitamin. Round up ready soy had also stunted the growth of rats in Monsanto's study but had not effected cattle although it had increased the fat content of their milk. The authors do not maintain that modified soy is a hazard but that the FDA accepting such significant differences as being *substantially equivalent* illustrates the need for more rigorous testing, and preferably not by the biotech industries themselves.

The value of current independent studies is problematic as, due to restrictive End-user agreements, researchers are forbidden by law from publishing independent research in peer reviewed journals without the approval of the agritech companies. Cornell University's Elson Shields, the spokesperson for a group of scientists who oppose this practice, submitted a statement to the United States Environmental Protection Agency (EPA) protesting that "as a result of restrictive access, no truly independent research can be legally conducted on many critical questions regarding the technology". Scientific American noted that several studies that were initially approved by seed companies were later blocked from publication when they returned "unflattering" results. While recognising that seed companies' intellectual property rights need to be protected, Scientific American calls the practice dangerous and has called for the restrictions on research in the End-user agreements to be lifted immediately and for the EPA to require, as a condition of approval, that independent researchers have unfettered access to GM products for testing.

The Welsh pressure group GM Free Cymru argues that governments should use independent studies rather than industry studies to assess crop safety. GM Free Cymru has also stated that independently funded researcher, Professor Bela Darvas of Debrecen University was refused Mon 863 Bt corn to use in his studies after previously publishing that a different variety of Monsanto corn was lethal to two Hungarian protected insect species and an insect classified as a rare.

Allergenicity

Worldwide, reports of allergies to all kinds of foods, particularly nuts, fish and shellfish, seem to be increasing, but it is not known if this reflects a genuine change in the risk of allergy, or an increased awareness of food allergies by the public. Some environmental organizations, such as the European Green Party and Greenpeace, have suggested that GM food might trigger food allergies, although other environmentalists have implicated causes as diverse as the greenhouse effect increasing pollen levels, greater exposure to synthetic chemicals, cleaner lifestyles, or more mold in buildings. A 2005 review in the journal *Allergy* of the results from allergen testing of current GM foods stated that "no biotech proteins in foods have been documented to cause allergic reactions".

A well-known case of a GM plant that did not reach the market due to it producing an allergic reaction was a new form of soybean intended for animal feed. The allergen was transferred unintentionally from the Brazil nut into genetically engineered soybeans, in a bid to improve soybean nutritional quality for animal feed use. This new protein increased the levels in the GM soybean of the natural essential amino acid methionine, which is commonly added to poultry feed. Investigation of the GM soybeans revealed that they produced immune reactions in people with Brazil nut allergies, since the methionine rich protein chosen by Pioneer Hi-Bred happened to be a major source of Brazil nut allergy. Although this soybean strain was not developed as a human food, Pioneer Hi-Bred discontinued further development of the GM soybean, due to the difficulty in ensuring that none of these soybeans entered the human food chain.

In November 2005 a pest-resistant field pea developed by the Australian CSIRO for use as a pasture crop was shown to cause an allergic reaction in mice. Work on this variety was immediately halted. The protein added to the pea did not cause the reaction in humans or mice in isolation, but when it was expressed in the pea, it exhibited a subtly different structure which may have caused the allergic reaction. The immunologist who tested the pea noted that crops need to be evaluated case-by-case, and a Greenpeace activist commented that to his knowledge, no countries required feeding studies for approval of genetically modified foods.

Plant scientist Maarten J Chrispeels has made these comments about this example: The recent Prescott et al. paper in JFAC contains a very interesting study on the immunogenicity of amylase [starch digestion enzyme] inhibitor in its native form (isolated from beans) and expressed as a transgene in peas. First of all, amylase inhibitor is a food protein, but also a "toxic" protein because it inhibits our digestive amylases. This is one of the reasons you have to cook your beans! (The other toxic bean protein is phytohemagglutinin and it is much more toxic).

This particular amylase inhibitor is found in the common bean (other species have other amylase inhibitors). Even though it is a food protein, it is unlikely ever to be used for genetic engineering of human foods because it inhibits our amylases. What the results show is that the protein, when synthesized in pea cotyledons has a different immunogenicity than when it is isolated from bean cotyledons (the native form). This is somewhat surprising but may be related to the presence of slightly different carbohydrate chains.

These cases of products that failed safety testing can either be viewed as evidence that genetic modification can produce unexpected and dangerous changes in foods, or alternatively that the current tests are effective at identifying any safety problems before foods come on the market.

Genetic modification can also be used to remove allergens from foods, which may, for example, allow the production of soy products that would pose a smaller risk of food allergies than standard soybeans. A hypo-allergenic strain of soybean was tested in 2003 and shown to lack the major allergen that is found in the beans. A similar approach has been tried in ryegrass, which produces pollen that is a major cause of hayfever: here a fertile GM grass was produced that lacked the main pollen allergen, demonstrating that the production of hypoallergenic grass is also possible.

Environmental Risks and Benefits

The large scale growth of GM plants may have both positive and negative effects on the environment. These may be both direct effects, on organisms that feed on or interact with the crops, or wider effects on food chains produced by increases or decreases in the numbers of other organisms. As an example of benefits, insect-resistant Bt-expressing crops will reduce the number of pest insects feeding on these plants, but as there are fewer pests, farmers do not have to apply as much insecticide, which in turn tends to increase the number of non-pest insects in these fields. A 2006 study of the global impact of GM crops, published by the UK consultancy PG Economics and funded by Illinois-Missouri Biotechnology Alliance, concluded that globally, the technology reduced pesticide spraying by 286,000 tons in 2006, decreasing the environmental impact of herbicides and pesticides by 15%.

By reducing the amount of ploughing needed, GM technology led to reductions of greenhouse gases from soil equivalent to removing 6.56 million cars from the roads. However, a 2009 study published by the Organic Center stated that the use of genetically engineered corn, soybean, and cotton increased the use of herbicides by 383 million pounds (191,500 tons), and pesticide use by 318.4 million pounds (159,200 tons). As an example of a concern about environmental risk, a lab at Cornell University published an article which caused worry in the US that Bt-corn pollen might affect the monarch butterfly. However this concern was disproved by six comprehensive articles in the Proceedings of the National Academy of Sciences in 2001. Monarch populations increased, despite increased Bt corn probably due to reduced pesticide use. Other possible effects might come from the spread of genes from modified plants to unmodified relatives, which might produce species of weeds resistant to herbicides. In some areas of the US "superweeds" have evolved naturally, these weeds are resistant to herbicides and have forced farmers to return to traditional crop management practices.

There has been controversy over the results of a farm-scale trial in the United Kingdom comparing the impact of GM crops and conventional crops on farmland biodiversity. Some claimed that the results showed that GM crops had a significant negative impact on wildlife. They pointed out that the studies showed that using herbicide resistant GM crops allowed better weed control and that under such conditions there were fewer weeds and fewer weed seeds. This result was then extrapolated to suggest that GM crops would have significant impact on the wildlife that might rely on farm weeds. The President of the Royal Society, the body that had carried out the trials, stated that "To generalize and declare 'all GM is bad' or 'all GM is good' for the environment as a result of these experiments is a gross over-simplification", arguing that although the trials showed that the combination of some GM crops with long-lasting herbicides were bad for biodiversity, using other GM crops without these herbicides increased biodiversity.

In July 2005 British scientists showed that transfer of a herbicide-resistance gene from GM oilseed rape to a wild cousin, charlock, and wild turnips was possible.

Many agricultural scientists and food policy specialists view GM crops as an important element in sustainable food security and environmental management. This point of view is summarized in the ABIC Manifesto:

On our planet, 18% of the land mass is used for agricultural production. This fraction cannot be increased substantially. It is absolutely essential that the yield per unit of land increases beyond current levels given that: The human population is still growing, and will reach about nine billion by 2040;70,000 km² of agricultural land (equivalent to 60% of the German agricultural area) are lost annually to growth of cities and other non-agricultural uses; Consumer diets in developing countries are increasingly changing from plant-based proteins to animal protein, a trend that requires a greater amount of crop-based feeds.

Other scientists, such as Dr. Charles Benbrook, argue that improvement of global food security is hardly being addressed by genetic research and that a lack of yield is often not caused by insufficient genetic resources. Regarding the issues of intellectual property and patent law, an international report from the year 2000 states:

If the rights to these tools are strongly and universally enforced - and not extensively licensed or provided pro bono in the developing world - then the potential applications of GM technologies described previously are unlikely to benefit the less developed nations of the world for a long time (ie until after the restrictions conveyed by these rights have expired).

Issues with Bt Maize

A well publicized claim associated with Bt crops (or transgenic maize) was the concern that pollen from Bt maize might kill the monarch butterfly. This report was puzzling because the pollen from most maize hybrids contains much lower levels of Bt than the rest of the plant and led to multiple follow-up studies. One possible issue revealed in these studies is the possibility that the initial study was flawed; based on the way the pollen was collected, in that they collected and fed non-toxic pollen that was mixed with anther walls that did contain Bt toxin. A collaborative research exercise was carried out over two years by several groups of scientists in the US and Canada, looking at the effects of Bt pollen in both the field and the laboratory. This resulted in a risk assessment that concluded that any risk posed by the corn to butterfly populations under real-world conditions was negligible. The USDA has stated that the weight of the evidence is that Bt crops do not pose a risk to the monarch butterfly. An independent 2002 review of the scientific literature concluded that "the commercial large-scale cultivation of current Bt–maize hybrids did not pose a significant risk to the monarch population" and noted that despite large-scale planting of GM crops, the butterfly's population is increasing.

In 2007 Andreas Lang, Éva Lauber and Béla Darvas criticized these studies, arguing that there can be a great difference in the effects between the acute exposure tested for and chronic exposure. Moreover, they stated that the "worst case conditions" performed were not in fact worst case scenarios, as laboratory conditions with ample food supply and a favorable climate ensure healthy subjects. They instead believe that in the wild, low temperatures, rain and parasites and disease might exacerbate a Bt effect on butterfly larvae. Their own experiments suggested that some butterfly species were negatively affected by such chronic exposure. Jörg Romeis, who conducted the original studies, replied that if species of Butterfly are affected as Darvas claims that a "more comprehensive assessment will be needed and, depending on the degree and nature of concern, this may extend to field testing".

A 2001 report in *Nature* presented evidence that Bt maize was cross-breeding with unmodified maize in Mexico, although the data in this paper was later described as originating from an artifact and *Nature* stated that "the evidence available is not sufficient to justify the publication of the original paper". A subsequent large-scale study, in 2005, failed to find any evidence of contamination in Oaxaca. However, other authors have stated that they also found evidence of cross-breeding between natural maize and transgenic maize.

There is also a risk that for example, transgenic maize will crossbreed with wild grass variants, and that the Bt-gene will end up in a natural environment, retaining its toxicity. An event like this would have ecological implications. However, there is no evidence of crossbreeding between maize and wild grasses.

In 2009 it was reported that 82,000 hectares (200,000 acres) of Bt corn in South Africa failed to produce seeds. Monsanto claimed average yield was reduced by 25% in those fields affected, it compensated the farmers concerned and the corn varieties were affected by a mistake made in the seed breeding process. Marian Mayet an environmental activitist and director of the Africa-centre for biosecurity in Johannesburg called for a government investigation and asserted that the biotechnology was at fault, "*You cannot make a 'mistake' with three different varieties of corn*". In 2009 South African farmers planted 1,900,000 hectares (4,700,000 acres) of GM maize (73% of the total crop).

As of 2007, a phenomenon called Colony Collapse Disorder (CCD) was noticed in bee hives all over North America, and elsewhere. Although it is not certain if this is a new phenomenon, initial ideas on the possible causes ranged from poor nutrition, infections, parasites and pesticide use. More unusual speculations included radio waves from cellphone base stations, climate change, and the use of transgenic crops containing Bt. The Mid-Atlantic Apiculture Research and Extension Consortium published a report on 2007-03-27 that found no evidence that pollen from Bt crops is adversely affecting bees. Several researchers in the US have since attributed CCD to the spread of a new virus called Israeli acute paralysis virus, although other parasites have also been implicated.

Legal Issues in the US

Alfalfa: On 21 June 2010, the US Supreme Court in *Monsanto v. Geertson Seed Farms* issued its first ruling about GM crops. This case relates to Roundup Ready alfalfa and dates back to 2006, when a coalition of groups led by the Center for Food Safety raised concerns about environmental impacts that the United States Department of

Agriculture (USDA) allegedly failed to address before approving the planting of Roundup Ready alfalfa. In response, in 2007, Judge Charles Breyer of the California Northern District Court ruled that USDA was in error when it approved the planting of Roundup Ready alfalfa. In June 2009, a divided three-judge panel on the 9th U.S. Circuit Court of Appeals upheld Judge Charles Breyer's decision, and Monsanto appealed to the Supreme Court. The impact of the current US Supreme Court ruling is somewhat unclear, with both sides appearing to claim victory.

Sugar Beets: There is also a somewhat similar case involving sugar beets before the same California Northern District Court. This is a case involving Monsanto's breed of pesticide-resistant sugar beets. While Judge Jeffrey S. White (issuing his ruling in the spring of 2010) allowed the planting of GM sugar beets to continue, he also warned that this may be blocked in the future while an environmental review is taking place.

Finally, on 13 August 2010, Judge Jeffrey S. White ordered the halt to the planting of the genetically modified sugar beets in the US. He indicated that "the Agriculture Department had not adequately assessed the environmental consequences before approving them for commercial cultivation." In 2010, before the ruling, 95% of the sugar beet grown in the US was GM. About half the sugar supply in the US came from sugar beet.

Control of the Market

Patent holder companies use their control of their own GMO to corner the market and gain profit. However, other companies often compete for the little market share available to GM foods worldwide. Detractors such as Greenpeace say that patent rights give corporations a dangerous amount of control over their product while corporations claim that they need product control in order to prevent seed piracy, fulfill financial obligations to shareholders and invest in further GM development. Governments have also sought to protect their commercial interests through punitive measures against countries resisting GM foods on ethical or scientific grounds: after moves in France to ban a Monsanto GM corn variety, the US embassy recommended that 'we calibrate a target retaliation list that causes some pain across the EU'.

Controversial Cases

Pusztai Potato Controversy: A large media event occurred in 1998 when scientist Árpád Pusztai, who was considered by some to be a world expert on plant lectins, reported in a television interview that he had found that rats fed potatoes genetically modified by the English biotech company *Cambridge Agricultural Genetics* to contain lectin, a natural insecticide in snowdrop plants, had stunted growth and immune system damage. Initially the Rowett Institute's director, Philip James, congratulated Pusztai after the interview, but his attitude changed after confusion arose over whether the lectin was from a jackbean or a snowdrop (the jackbean lectin is poisonous to the intestines). A press release from the Rowett Institute stated that the lectin had come from a jackbean, but no experiments were made using lectins from the jackbean; in the confusion, it was alleged that Pusztai hadn't actually carried out the experiments. James suspended Pusztai and used misconduct procedures to seize the data.

His annual contract was not renewed and both Pusztai and his wife were banned from speaking publicly. In October 1998 the Rowett Institute published an audit criticizing Pusztai's results, which, along with Pusztai's raw data, was sent to six anonymous reviewers who criticized Pusztai's work. Pusztai responded that the raw data was "never intended for publication under intense scrutiny". Pusztai sent the audit report and his rebuttal to scientists who requested it, and in February 1999, twenty-one European and American scientists, mostly friends or acquaintances of Pusztai, orchestrated by the environmental group Friends of the Earth, released a memo supporting Pusztai. Stanley Ewen, who worked with Pusztai, conducted a followup study supporting Pusztai's work and presented the work to a lectin meeting in Sweden.

In October 1999 Pusztai's research was published in the journal The Lancet. Because of the controversial nature of his research the data in this paper, co-authored by Stanley Ewen, was seen by a total of six reviewers when presented for peer review; five of these reviewers judged the work acceptable, although one of the five "deemed the study flawed but favored publication to avoid suspicions of a conspiracy against Pusztai and to give colleagues a chance to see the data for themselves". The paper did not mention stunted growth or immunity issues, but reported that rats fed on potatoes genetically modified with the snowdrop lectin had "thickening in the mucosal lining of their colon and their jejunum" when compared with rats fed on non modified potatoes. Three Dutch scientists criticized the study on the grounds that the unmodified potatoes were not a fair control diet, and that any rats fed only on potatoes will suffer from a protein deficiency; Pusztai responded to these criticisms by stating that the protein and energy were comparable, and that "a sample size of six is perfectly normal in studies like this".

The genetically modified potatoes were later destroyed by *Cambridge Agricultural Genetics* as were all details of their modification.

Bioequivalence Study of a Corn Cultivar

A controversy arose around biotech company Monsanto's data on a 90-Day Rat Feeding Study on the MON863 strain of GM corn. In May 2005, critics of GM foods pointed to differences in kidney size and blood composition found in this study, suggesting that the observed differences raises questions about the regulatory concept of substantial equivalence. Anti-GM campaigner Jeffrey M. Smith, writing in Biophile Magazine, quoting comments from Pusztai and Seralini,has stated that nutritional studies typically use young, fast-growing animals with starting weights not varying by more than 2% from the average whereas Monsanto's research design used a mix of young and old animals with starting weights ranging from 198.4 to 259.8 grams. Seralini and two other authors published a study of these data, funded by Greenpeace, in 2007 making similar points.

The raising of this issue prompted the European Food Safety Authority (EFSA) to reexamine the safety data on this strain of corn. The EFSA concluded that the observed small numerical decrease in rat kidney weights were not biologically meaningful, and the weights were well within the normal range of kidney weights for control animals. There were no corresponding microscopic findings in the relevant organ systems, and they stated that all blood chemistry and organ weight values fell within the "normal range of historical control values" for rats. In addition the EFSA review stated that the statistical methods used by Séralini et al. in the analysis of the data were incorrect. The European Commission has approved the ÌÌÍ863 corn for animal and human consumption. Food Standards Australia New Zealand reviewed the 2007 Seralini et al study and concluded that "...all of the statistical differences between rats fed MON 863 corn and control rats are attributable to normal biological variation."

Greenpeace stated in a 2007 press release that Séralini et al. had completed a similar analysis of the NK603 strain of corn and came to similar conclusions as they did in their previous study. Séralini et al included this in a re-analysis of three existing rat feeding studies published in 2009.

The European Food Safety Authority reviewed the 2009 Seralini et al paper and concluded that the author's claims were not supported by the data in their paper. They noted that many of their fundamental statistical criticisms of the 2007 paper also applied to the 2009 paper. There was no new information that would change the EFSA's conclusions that the three GM maize types were safe for human, animal health and the environment The French High Council of Biotechnologies Scientific Committee (HCB) also reviewed the 2009 study and concluded that it "..presents no admissible scientific element likely to ascribe any haematological, hepatic or renal toxicity to the three re-analysed GMOs." The HCB also questioned the author's independence. Food Standards Australia New Zealand concluded that the results from the 2009 Séralini et al. study were due to chance alone.

Contamination Issues

In the 1990s genetically modified Flax tolerant to herbicide residues in soil was developed by the Crop Development Centre (CDC) at the University of Saskatchewan in Canada. Named Flax variety FP967, but commonly called *CDC Triffid*, research was controversially halted following protests from Canadian farmers who stood to lose up to 70% of their traditional export markets if it was introduced. GM Flax was deregistered, its sale was criminalized and in 2001 all modified seeds were destroyed. No modified crops had been planted and no seed had been sold but GM industry proponent Alan McHughen controversially passed out sample packets of seeds at presentations. In early September 2009, Flax imported into Germany was found to be contaminated with *CDC Triffid* causing the price of Canadian Flax to fall 32 percent. By mid November 35 countries reported contamination of imported Canadian Flax which has now been banned by the European Union. Canadian farmers are expected to be responsible for the cost of the cleanup and testing of future crops.

In 2000, Aventis StarLink corn, which had been approved only as animal feed due to concerns about possible allergic reactions in humans, was found contaminating corn products in U.S. supermarkets. An episode involving Taco Bell taco shells was particularly well publicized which resulted in sales of StarLink seed being discontinued. The registration for the Starlink varieties was voluntarily withdrawn by Aventis in October 2000. Aid sent by the UN and the US to Central African nations was also found to be contaminated with StarLink corn and the aid was rejected. The US corn supply has been monitored for

Starlink Bt proteins since 2001 and no positive samples have been found since 2004.

Public Perception

Research by the Pew Initiative on Food and Biotechnology has shown that in 2005 Americans' knowledge of genetically modified foods and animals continues to remain low, and their opinions reflect that they are particularly uncomfortable with animal cloning. In one instance of consumer confusion, DNA Plant Technology's Fish tomato transgenic organism was conflated with Calgene's Flavr Savr transgenic food product. The Pew survey also showed that despite continuing concerns about GM foods, American consumers do not support banning new uses of the technology, but rather seek an active role from regulators to ensure that new products are safe.

Only 2% of Britons were said to be "happy to eat GM foods", and more than half of Britons were against GM foods being available to the public, according to a 2003 study. However a 2009 review article of European consumer polls concluded that opposition to GMOs in Europe has been gradually decreasing. Approximately half of European consumers accepted gene technology, particularly when benefits for consumers and for the environment could be linked to GMO products. 80 % of respondents did not cite the application of GMOs in agriculture as a significant environmental problem. Many consumers seem unafraid of health risks from GMO products and most European consumers did not actively avoid GMO products while shopping.

In Australia, GM foods that have novel DNA, novel protein, altered characteristics or has to be cooked or prepared in a different way compared to the conventional food have, since December 2001, had to be identified on food labels. However, multiple surveys have shown that while 45% of the public will accept GM foods, some 93% demand all genetically modified foods be labelled as such. A 2007 survey by the Food Standards Australia and New Zealand found that 27% of Australians looked at the label to see if it contained GM material when purchasing a grocery product for the first time. Labelling legislation has been introduced and rejected several times since 1996 on the grounds of "restraint of trade" due to the cost of labelling. The controversy erupted again in 2009 when Graincorp, the nations largest grain handler, announced it would mix GM Canola with its unmodified grain. Traditional growers, who largely rely on GM-free markets, had been told they would need to pay to have their produce certified GM free. Graincorp reversed its decision the same year. Critics such as Greenpeace and the Gene Ethics Network have renewed calls for more labelling.

Opponents of genetically modified food often refer to it as "Frankenfood", after Mary Shelley's character Frankenstein and the monster he creates, in her novel of the same name. The term was coined in 1992 by Paul Lewis, an English professor at Boston College who used the word in a letter he wrote to the *New York Times* in response to the decision of the US Food and Drug Administration to allow companies to market genetically modified food. The term "Frankenfood" has become a battle cry of the European side in the US-EU agricultural trade war.

Critics have protested in regards to the appointment of pro GM lobbyists to senior positions in the FDA. Michael R. Taylor has been appointed as a senior adviser to the FDA on food safety and Dennis Wolff is expected to take up the position of Under-Secretary of the newly created Agriculture for Food Safety. Taylor is a former Monsanto lobbyist credited as being responsible for the implementation of "substantial equivalence" in place of food safety studies and for his advocacy that resulted in the Delaney clause that prohibited the inclusion of "any chemical additive found to induce cancer in man.. or animals" in processed foods being amended in 1996 to allow the inclusion of pesticides in GMOs.

Wolff is the Pennsylvania Secretary of Agriculture who successfully lobbied to ban organic farmers from labelling their products as being GM free and was a proponent of the "ACRE" initiative which gave the Pennsylvania state attorney general's office the authority to sue municipalities that banned GMOs. Several anti-GMO organisations have organised petitions demanding Taylor's resignation and opposing Wolff's appointment and also conducted letter writing campaigns protesting the conflict of interest.

Genetically Modified Organism

A genetically modified organism (GMO) or genetically engineered organism (GEO) is an organism whose genetic material has been altered using genetic engineering techniques. These techniques, generally known as recombinant DNA technology, use DNA molecules from different sources, which are combined into one molecule to create a new set of genes. This DNA is then transferred into an organism, giving it modified or novel genes. Transgenic organisms, a subset of GMOs, are organisms which have inserted DNA that originated in a different species.

Production

Genetic modification involves the insertion or deletion of genes. When genes are inserted, they usually come from a different species, which is a form of horizontal gene transfer. In nature this can occur when exogenous DNA penetrates the cell membrane for any reason. To do this artificially may require attaching the genes to a virus or just physically inserting the extra DNA into the nucleus of the intended host with a very small syringe, or with very small particles fired from a gene gun. However, other methods exploit natural forms of gene transfer, such as the ability of *Agrobacterium* to transfer genetic material to plants, or the ability of lentiviruses to transfer genes to animal cells.

History

The general principle of producing a GMO is to add new genetic material into an organism's genome. This is called genetic engineering and was made possible through the discovery of DNA and the creation of the first recombinant bacteria in 1973; an existing bacterium *E. coli* expressing an exogenic Salmonella gene. This led to concerns in the scientific community about potential risks from genetic engineering, which were first discussed in depth at the Asilomar Conference in 1975. One of the main recommendations from this meeting was that government oversight of recombinant DNA research should be established until the technology was deemed safe.

Herbert Boyer then founded the first company to use recombinant DNA technology, Genentech, and in 1978 the company announced creation of an *E. coli* strain producing the human protein insulin. In 1986, field tests of bacteria genetically engineered to protect plants from frost damage (ice-minus bacteria) at a small biotechnology company called Advanced Genetic Sciences of Oakland, California, were repeatedly delayed by opponents of biotechnology. In the same year, a proposed field test of a microbe genetically engineered for a pest resistance protein by Monsanto Company was dropped. In the late 1980s and early 1990s guidance on assessing the safety of genetically engineered plants and food emerged from organizations including the FAO and WHO. Small scale experimental plantings of genetically modified (GM) plants began in Canada and the U.S. in the late 1980s. The first approvals for large scale, commercial cultivation came in the mid 1990s. Since that time, adoption of GM plants by farmers has increased annually.

Uses

GMOs are used in biological and medical research, production of pharmaceutical drugs, experimental medicine (e.g. gene therapy), and agriculture (e.g. golden rice). The term "genetically modified organism" does not always imply, but can include, targeted insertions of genes from one species into another. For example, a gene from a jellyfish, encoding a fluorescent protein called GFP, can be physically linked and thus co-expressed with mammalian genes to identify the location of the protein encoded by the GFP-tagged gene in the mammalian cell. Such methods are useful tools for biologists in many areas of research, including those who study the mechanisms of human and other diseases or fundamental biological processes in eukaryotic or prokaryotic cells.

To date the most controversial but also the most widely adopted application of GMO technology is patent-protected food crops which are resistant to commercial herbicides or are able to produce pesticidal proteins from within the plant, or *stacked trait* seeds, which do both. The largest share of the GMO crops planted globally are owned by the US firm Monsanto.

In 2007, Monsanto's trait technologies were planted on 246 million acres (1,000,000 km^2) throughout the world, a growth of 13 percent from 2006. However, patents on the first Monsanto products to enter the marketplace will begin to expire in 2014, democratizing Monsanto products. In addition, a 2007 report from the European Joint Research Commission predicts that by 2015, more than 40 per cent of new GM plants entering the global marketplace will have been developed in Asia.

In the corn market, Monsanto's triple-stack corn—which combines Roundup Ready 2 weed control technology with YieldGard Corn Borer and YieldGard Rootworm insect control—is the market leader in the United States. U.S. corn farmers planted more than 32 million acres (130,000 km^2) of triple-stack corn in 2008, and it is estimated the product could be planted on 56 million acres (230,000 km^2) in 2014–2015. In the cotton market, Bollgard II with Roundup Ready Flex was planted on approximately 5 million acres (20,000 km^2) of U.S. cotton in 2008.

According to the International Service for the Acquisition of Agri-Biotech Applications (ISAAA), of the approximately 14 million farmers who grew biotech crops in 2009, some 90% were resource-poor farmers in developing countries. These include some 7 million farmers in the cotton-growing areas of China, an estimated 5.6 million small farmers in India (Bt cotton), 250,000 in the Philippines, South Africa (biotech cotton, maize and soybeans often grown by subsistence women farmers) and the other twelve developing countries which grew biotech crops in 2009. 10 million more small and resource-poor farmers may have been secondary beneficiaries of Bt cotton in China. The global commercial value of biotech crops grown in 2008 was estimated to be US$130 billion.

In the United States, the United States Department of Agriculture (USDA) reports on the total area of GMO varieties planted. According to National Agricultural Statistics Service, the states published in these tables represent 81–86 percent of all corn planted area, 88–90 percent of all soybean planted area, and 81–93 percent of all upland cotton planted area (depending on the year).

USDA does not collect data for global area. Estimates are produced by the International Service for the Acquisition of Agri-biotech Applications (ISAAA) and can be found in the report, "Global Status of Commercialized Transgenic Crops: 2007".

Transgenic animals are also becoming useful commercially. On February 6, 2009 the U.S. Food and Drug Administration approved the first human biological drug produced from such an animal, a goat. The drug, ATryn, is an anticoagulant which reduces the probability of blood clots during surgery or childbirth. It is extracted from the goat's milk.

Detection

Testing on GMOs in food and feed is routinely done by molecular techniques like DNA microarrays or qPCR. The test can be based on screening elements (like p35S, tNos, pat, or bar) or event specific markers for the official GMOs (like Mon810, Bt11, or GT73). The array-based method combines multiplex PCR and array technology to screen samples for different potential GMOs, combining different approaches (screening elements, plant-specific markers, and event-specific markers). The qPCR is used to detect specific GMO events by usage of specific primers for screening elements or event-specific markers.

To avoid any kind of false positive or false negative testing outcome, comprehensive controls for every step of the process is mandatory. A CaMV check is important to avoid false positive outcomes based on virus contamination of the sample.

Transgenic Microbes

Bacteria were the first organisms to be modified in the laboratory, due to their simple genetics. These organisms are now used for several purposes, and are particularly important in producing large amounts of pure human proteins for use in medicine.

Genetically modified bacteria are used to produce the protein insulin to treat diabetes. Similar bacteria have been used to produce clotting factors to treat haemophilia, and human growth hormone to treat various forms of dwarfism.

Transgenic Animals

Transgenic animals are used as experimental models to perform phenotypic and for testing in biomedical research. Other applications include the production of human hormones such as insulin.

Fruit Flies

In biological research, transgenic fruit flies (*Drosophila melanogaster*) are model organisms used to study the effects of genetic changes on development. Fruit flies are often preferred over other animals due to their short life cycle, low maintenance requirements, and relatively simple genome compared to many vertebrates.

Mammals

Genetically modified mammals are an important category of genetically modified organisms. Transgenic mice are often used to study cellular and tissue-specific responses to disease.

In 1999, scientists at the University of Guelph in Ontario, Canada created the genetically engineered Enviropig. The Enviropig excretes from 30 to 70.7% less phosphorus in manure depending upon the age and diet. In February 2010, Environment Canada determined that Enviropigs are in compliance with the Canadian Environmental Protection Act and can be produced outside of the research context in controlled facilities where they are segregated from other animals.

In 2009, scientists in Japan announced that they had successfully transferred a gene into a primate species (marmosets) and produced a stable line of breeding transgenic primates for the first time.

Cnidarians

Cnidarians such as *Hydra* and the sea anemone Nematostella vectensis have become attractive model organisms to study the evolution of immunity and certain developmental processes. An important technical breakthrough was the development of procedures for generation of stably transgenic hydras and sea anemones by embryo microinjection.

Fish

Genetically modified fish have promoters driving an over-production of "all fish" growth hormone. This resulted in dramatic growth enhancement in several species, including salmonids, carps and tilapias.

Gene Therapy

Gene therapy, uses genetically modified viruses to deliver genes that can cure disease into human cells. Although gene therapy is still relatively new, it has had some successes. It has been used to treat genetic disorders such as severe combined immunodeficiency, and treatments are being developed for a range of other currently incurable diseases, such as cystic fibrosis, sickle cell anemia, and muscular dystrophy. Current gene therapy technology only targets the non-reproductive cells meaning that any changes introduced by the treatment can not be transmitted to the next generation. Gene therapy targeting the reproductive cells—so-called "Germ line Gene Therapy"—is very controversial and is unlikely to be developed in the near future.

Transgenic Plants

Transgenic plants have been engineered to possess several desirable traits, such as resistance to pests, herbicides, or harsh environmental conditions, improved product shelf life, and increased nutritional value. Since the first commercial cultivation of genetically modified plants in 1996, they have been modified to be tolerant to the herbicides glufosinate and glyphosate, to be resistant to virus damage as in Ringspot virus-resistant GM papaya, grown in Hawaii, and to produce the Bt toxin, an insecticide that is non-toxic to mammals.

Most GM crops grown today have been modified with "input traits", which provide benefits mainly to farmers. The GM oilseed crops on the market today offer improved oil profiles for processing or healthier edible oils. The GM crops in development offer a wider array of environmental and consumer benefits such as nutritional enhancement, drought and stress tolerance.

GM plants are being developed by both private companies and public research institutions such as CIMMYT, the International Maize and Wheat Improvement Centre. Other examples include a genetically modified sweet potato, enhanced with protein and other nutrients, while golden rice, developed by the International Rice Research Institute (IRRI), has been discussed as a possible cure for Vitamin A deficiency.

The coexistence of GM plants with conventional and organic crops has raised significant concern in many European countries. Due to relatively high demand from European consumers for the freedom of choice between GM and non-GM foods, EU regulations require measures to avoid mixing of foods and feed produced from GM crops and conventional or organic crops. European research programs such as Co-Extra, Transcontainer, and SIGMEA are investigating appropriate tools and rules. At the field level, biological containment methods include isolation distance and pollen barriers. Such measures are generally not used in North America because they are very costly and there are no safety-related reasons to employ them.

Cisgenic Plants

Cisgenesis, sometimes also called Intragenesis, is a product designation for a category of genetically engineered plants. A variety of classification schemes have been proposed, that order genetically modified organisms based on the nature of introduced genotypical changes rather than the process of genetic engineering.

While some genetically modified plants are developed by the introduction of a gene originating from distant, sexually incompatible species into the host genome, cisgenic plants contain genes which have been isolated either directly from the host species or from sexually compatible species. The new genes are introduced using recombinant DNA methods and gene transfer. Some scientists hope that the approval process of cisgenic plants might be simpler than that of proper transgenics, but it remains to be seen.

Controversy

Biological Process: The use of genetically modified organisms has sparked significant controversy in many areas. Some groups or individuals see the generation and use of GMO as intolerable meddling with biological states or processes that have naturally evolved over long periods of time, while others are concerned about the limitations of modern science to fully comprehend all of the potential negative ramifications of genetic manipulation. Other people see this as a continuation in the role humanity has occupied for thousands of years, modifying the genetics of crops.

Foodchain

The safety of GMOs in the foodchain has been questioned by some environmental groups, with concerns such as the possibilities that GMOs could introduce new allergens into foods, or contribute to the spread of antibiotic resistance.

All studies published to date have shown no adverse health effects resulting from humans eating genetically modified foods, environmental groups still discourage consumption in many countries, claiming that GM foods are unnatural and therefore unsafe. Such concerns have led to the adoption of laws and regulations that require safety testing of any new organism produced for human consumption.

GMOs' proponents note that because of the safety testing requirements imposed on GM foods, the risk of introducing a plant variety with a new allergene or toxin using genetic modification is much smaller than using traditional breeding processes. An example of an allergenic plant created using traditional breeding is the kiwi. One article calculated that the marketing of GM salmon could reduce the cost of salmon by half, thus increasing salmon consumption and preventing 1,400 deaths from heart attack a year in the United States.

Trade in Europe and Africa

In response to negative public opinion, Monsanto announced its decision to remove their seed cereal business from Europe, and environmentalists crashed a World Trade Organization conference in Cancun that promoted GM foods and was sponsored by Committee for a Constructive Tomorrow (CFACT). Some African nations have refused emergency food aid from developed countries, fearing that the food is unsafe.

During a conference in the Ethiopian capital of Addis Ababa, Kingsley Amoako, Executive Secretary of the United Nations Economic Commission for Africa (UNECA), encouraged African nations to accept genetically modified food and expressed dissatisfaction in the public's negative opinion of biotechnology.

Agricultural Surpluses

Patrick Mulvany, Chairman of the UK Food Group, accused some governments, especially the Bush administration, of using GM food aid as a way to dispose of unwanted agricultural surpluses. The UN blamed food companies and accused them of violating human rights, calling on governments to regulate these profit-driven firms. It is widely believed that the acceptance of biotechnology and genetically modified foods will also benefit rich research companies and could possibly benefit them more than consumers in underdeveloped nations.

Labelling

While some groups advocate the complete prohibition of GMOs, others call for mandatory labelling of genetically modified food or other products. Other controversies include the definition of patent and property pertaining to products of genetic engineering. According to the documentary Food, Inc. efforts to introduce labelling of GMOs has repetedly met resistance from lobbyists and politicians affiliated with companies like Monsanto.

Testing

Bruce Stutz's article, "Wanted: GM Seeds for Study," highlights a story of two dozen scientists who spoke out against the research restrictions put forth by companies producing genetically modified (GM) seeds such as DuPont, Monsanto, and Syngenta. In February 2009, after scientists warned the U.S. Environmental protection Agency (EPA) "that industry influence had made independent analyses of transgenic crops impossible," the American Seed Trade Association (ASTA) agreed that they "would allow researchers greater freedom to study the effects of GM food crops." This agreement left many scientists optimistic about the future, but there is little optimism as to whether this agreement has the ability to "alter what has been a research environment rife with obstruction and suspicion."

Impoverished Nations

Some groups believe that impoverished nations will not reap the benefits of biotechnology because they do not have easy access to these developments, cannot afford modern agricultural equipment, and certain aspects of the system revolving around intellectual property rights are unfair to "undeveloped countries". For example, The CGIAR (Consultative Group of International Agricultural Research) is an aid and research organization that has been working to achieve sustainable food security and decrease poverty in undeveloped countries since its formation in 1971. In an evaluation of CGIAR, the World Bank praised its efforts but suggested a shift to genetics research and productivity enhancement.

This plan has several obstacles such as patents, commercial licenses, and the difficulty that third world countries have in accessing the international collection of genetic resources and other intellectual property rights that would educate them about modern technology. The International Treaty on Plant Genetic Resources for Food and Agriculture has attempted to remedy this problem, but results have been inconsistent. As a result, "orphan crops", such as teff, millets, cowpeas, and indigenous plants, are important in the countries where they are grown, but receive little investment.

Private Investments

The development and implementation of policies designed to encourage private investments in research and marketing biotechnology that will meet the needs of poverty-stricken nations, increased research on other problems faced by poor nations, and joint efforts by the public and private sectors to ensure the efficient use of technology developed by industrialized nations have been suggested. In addition, industrialized nations have not tested GM technology on tropical plants, focusing on those that grow in temperate climates, even though undeveloped nations and the people that need the extra food live primarily in tropical climates. Some European scientists are concerned that political factors and ideology prevent unbiased assessment of GM technology in some EU countries, with a negative effect on the whole community.

Transgenic Organisms

Another important controversy is the possibility of unforeseen local and global effects as a result of transgenic organisms proliferating. The basic ethical issues involved in genetic research are discussed in the article on genetic engineering.

Some critics have raised the concern that conventionally-bred crop plants can be cross-pollinated (bred) from the pollen of modified plants. Pollen can be dispersed over large areas by wind, animals, and insects. In 2007, the U.S. Department of Agriculture fined Scotts Miracle-Gro $500,000 when modified genetic material from creeping bentgrass, a new golf-course grass Scotts had been testing, was found within close relatives of the same genus (*Agrostis*) as well as in native grasses up to 21 km (13 miles) away from the test sites, released when freshly cut grass was blown by the wind.

GM proponents point out that outcrossing, as this process is known, is not new. The same thing happens with any new open-pollinated crop variety—newly introduced traits can potentially cross out into neighboring crop plants of the same species and, in some cases, to closely related wild relatives. Defenders of GM technology point out that each GM crop is assessed on a case-by-case basis to determine if there is any risk associated with the outcrossing of the GM trait into wild plant populations.

The fact that a GM plant may outcross with a related wild relative is not, in itself, a risk unless such an occurrence has negative consequences. If, for example, an herbicide resistance trait was to cross into a wild relative of a crop plant it can be predicted that this would not have any consequences except in areas where herbicides are sprayed, such as a farm. In such a setting the farmer can manage this risk by rotating herbicides.

The European Union funds research programs such as Co-Extra, that investigate options and technologies on the coexistence of GM and conventional farming. This also includes research on biological containment strategies and other measures to prevent outcrossing and enable the implementation of coexistence.

If patented genes are outcrossed, even accidentally, to other commercial fields and a person deliberately selects the outcrossed plants for subsequent planting then the patent holder has the right to control the use of those crops. This was supported in Canadian law in the case of Monsanto Canada Inc. v. Schmeiser. However, F2 seed (next generation from the patented uniform crop) is highly variable as the crossed genes segregate. This makes it a much more variable crop and much less reliable for resistance or yield, so deliberate saving of seed is not practiced with patented crops. Instead new seed is bought each year, the farmer choosing a higher cost of seed to generate a much lower cost in insecticides/pesticides or for insurance of reliability and yield of the crop.

"Terminator" and "Traitor"

An often cited controversy is a "Technology Protection" technology dubbed 'Terminator'. This uncommercialized technology would allow the production of first generation crops that would not generate seeds in the second generation because the plants yield sterile seeds. The patent for this so-called "terminator" gene technology is owned by *Delta and Pine Land Company* and the United States Department of Agriculture. Delta and Pine Land was bought by Monsanto Company in August 2006. Similarly, the hypothetical trait-specific Genetic Use Restriction Technology, also known as 'Traitor' or 'T-gut', requires application of a chemical to genetically modified crops to reactivate engineered traits. This technology is intended both to limit the spread of genetically engineered plants, and to require farmers to pay yearly to reactivate the genetically engineered traits of their crops. Genetic Use Restriction Technology is under development by companies including Monsanto and AstraZeneca.

In addition to the commercial protection of proprietary technology in self-pollinating crops such as soybean (a generally contentious issue), another purpose of the terminator gene is to prevent the escape of genetically modified traits from cross-pollinating crops into wild-type species by sterilizing any resultant hybrids. Some environmentalist groups, while considering outcrossing of GM plants dangerous, felt the technology would prevent re-use of seed by farmers growing such terminator varieties in the developing world and was ostensibly a means to exercise patent claims.

However other environmental groups welcomed the terminator gene as a means of preventing GM crops from mixing with natural crops. Hybrid seeds were commonly used in the developed countries long before the introduction of GM crops. Hybrid seeds cannot be saved, so purchasing new seed every year is already a standard agricultural practice. There are technologies evolving which contain the transgene by biological means and still can provide fertile seeds using fertility restorer functions. Such methods are being developed by several EU research programs, among them Transcontainer and Co-Extra.

Governmental Support and Opposition

Australia: Several states of Australia had placed bans on planting GM food crops, beginning in 2003. However, in late 2007 the states of New South Wales and Victoria lifted their bans. Western Australia lifted their state's ban in December 2008, while South Australia continues its ban. Tasmania has extended its moratorium until November 2014. The state of Queensland has allowed the growing of GM crops since 1995 and has never had a GM ban.

Canada: Genetically modified crops have been widely adopted in Canada and have been grown since 1995. Nearly all of the canola grown in Canada is GM, as are significant proportions of corn and soybean. The Canadian regulatory system for biotechnology is science-based.

In general, biotechnology is well-accepted by Canadian consumers and farmers, with some exceptions. In 2005, a standing committee of the government of Prince Edward Island (PEI) in Canada assessed a proposal to ban the production of GMOs in the province. The ban was not passed. As of January 2008, the use of genetically modified crops on PEI was rapidly increasing. Mainland Canada is one of the world's largest producers of GM canola.

Japan: As of 2009, Japan has no commercial farming of any kinds of genetically modified food. Consumers have strongly resisted both imports and attempts to grow GMO in the country. Campaigns by consumer groups and environmental groups, such as Consumers Union of Japan and Greenpeace Japan, as well as local campaigns, have been very successful. In Hokkaido, a special bylaw has made it virtually impossible to grow GMOs, as the No! GMO Campaign collected over 200,000 signatures to oppose GMO farming. Consumers Union of Japan participated together with other Japanese NGOs at the Planet Diversity conference in Bonn, Germany on May 12–16, 2008, a global congress on the future of food and agriculture, with a demonstration to celebrate biodiversity, to oppose GMOs. "We don't only need networks between people, but between people and plants, and people and planet earth," noted Koketsu Michiyo from CUJ.

Cross-pollination has commonly occurred in Japan, as canola seed (rape seed) is imported from Canada. Around ports and the roads to major food oil companies, GE canola has now been found growing wild. Imported canola seeds have been found to be GMO varieties, including the Roundup Ready and Liberty Link types not grown in Japan. Activists and local groups, as well as the No! GMO Campaign and others, are alarmed that imported GMOs may harm the biodiversity and cause irreversible damage. A report from the Japanese National Institute for Environmental Studies (NIES) confirms that herbicide-resistant genetically engineered canola plants were identified in five of the six Japanese ports where samples were collected.

A number of Japanese groups have been making submissions to Western Australia's Review of the Genetically Modified Crops Free Areas Act 2003. These include the Seikatsu Club Consumers'

Cooperative Union and the Consumers Union of Japan. Seikatsu—an umbrella group of 29 Seikatsu Club Consumers' Co-Operatives—and its oil crushers Okamura Oil Mill Ltd and Yonezawa Oil Co. Ltd., all have non-GE canola policies. The groups stopped importing canola from Canada after the introduction of GE canola, when cross-pollination made it impossible to guarantee GE-free canola from Canada.

Pakistan: The government supports the use of hybrid seeds. However, Monsanto once tried to sell their hybrid seeds of such important crops as wheat and rice via the government. Even though yields would have increased, it would have made the Pakistani population dependent on the seeds of one company. The contract was never given.

New Zealand: In New Zealand, no genetically modified food is grown and no medicines containing live genetically-modified organisms have been approved for use. However, medicines manufactured using genetically modified organisms that do not contain live organisms have been approved for sale, and imported foods with genetically modified components are sold.

United States: In 2004, Mendocino County, California became the first county in the United States to ban the production of GMOs. The measure passed with a 57% majority. In California, Trinity and Marin counties have also imposed bans on GM crops, while ordinances to do so were unsuccessful in Butte, Lake, San Luis Obispo, Humboldt, and Sonoma counties. Supervisors in the agriculturally-rich counties of Fresno, Kern, Kings, Solano, Sutter, and Tulare have passed resolutions supporting the practice.

In 2007, with reference to US negotiations with the EU on agricultural biotechnology, US diplomatic cables recommended that 'we calibrate a target retaliation list that causes some pain across the EU'.

Zambia: The Zambian government has launched a campaign to educate and increase awareness of the benefits of biotechnology, including genetically modified crops, in order to change negative public opinion.

Other Africa: In 2010, after nine years of talks, the Common Market for Eastern and Southern Africa (COMESA) produced a draft policy on GM technology. This proposed policy was sent to all 19 national governments for consultation in September 2010.

Under the policy, a member country which wants to grow a new GM crop would inform COMESA who would have sufficient scientific expertise to make the decision as to whether the crop was safe for the environment and for humans. At the moment, few countries have the resources to make their own decisions. Once COMESA had made their decision, permission would be granted for the crop to be grown in all 19 member countries. Member countries would retain the power not to grow the crop in their own country if they wanted.

France: The cultivation of Monsanto's MON 810 corn was forbidden in France on February 9, of 2008. It was the only GMO authorized in France. The safeguard measure is taken as far as side effects on human health will be known. In 2010 Marion Guillou, president of the National Institute for Agronomical Research and one of France's top farm researcher, said she can no longer work on developing new GMOs due to widespread distrust and even hostility by European consumers.

Germany: Germany placed a ban on the cultivation and sale of GMO maize in April 2009.

Other European Countries

MON 810 (maize) was the first GMO crop to be cultivated in Europe. The initial lines of maize were approved in 1997 and, by 2009, 76,000 hectares of GM maize were grown in Spain (20% of Spain's maize production). Smaller amounts were produced in the Czech Republic, Slovakia, Portugal, Romania and Poland. However, in addition to France and Germany, other European countries that have placed bans on the cultivation and sale of GMOs include Austria, Hungary, Greece, and Luxembourg. Ireland has also banned GMO cultivation, and has instituted a voluntary label for GMO-free food products. Poland has also tried to institute a ban, with backlash from the European Commission. Bulgaria effectively banned cultivation of genetically modified organisms on March 18, 2010.

On 2 March 2010 a second species of GMO, a potato named Amflora, was approved for cultivation for industrial applications in the EU by the European Commission and was grown in Germany, Sweden and the Czech Republic that year. On 13 July 2010, the European Commission issued a recommendation that in future individual states in the EU should be able to ban the growing of specific GM crops that had been scientifically approved at the EU level. A ban could be justified on cultural, economic or ethical grounds.

The EU approval process for imports of GM crops and labelling of GM food products remained in place.

Marker Assisted Selection a Way to Select Suitable Offspring without Using Genetic Engineering

Marker assisted selection or marker aided selection (MAS) is a process whereby a marker (morphological, biochemical or one based on DNA/RNA variation) is used for indirect selection of a genetic determinant or determinants of a trait of interest (i.e. productivity, disease resistance, abiotic stress tolerance, and/or quality). This process is used in plant and animal breeding. Considerable developments in biotechnology have led plant breeders to develop more efficient selection systems to replace traditional phenotypic-pedigree-based selection systems.

Marker assisted selection (MAS) is indirect selection process where a trait of interest is selected not based on the trait itself but on a marker linked to it. For example if MAS is being used to select individuals with a disease, the level of disease is not quantified but rather a marker allele which is linked with disease is used to determine disease presence. The assumption is that linked allele associates with the gene and/or quantitative trait locus (QTL) of interest. MAS can be useful for traits that are difficult to measure, exhibit low heritability, and/or are expressed late in development.

Marker Types

A marker may be:

- Biological- Different pathogen races or insect biotypes based on host pathogen or host parasite interaction can be used as a marker since the genetic constitution of an organism can affect its susceptibility to pathogens or parasites.

- Morphological - First markers loci available that have obvious impact on morphology of plant. Genes that affect form, coloration, male sterility or resistance among others have been analysed in many plant species. Examples of this type of marker may include the presence or absence of awn, leaf sheath coloration, height, grain color, aroma of rice etc. In well-characterized crops like maize, tomato, pea, barley or wheat, tens or even hundreds of such genes have been assigned to different chromosomes.

- Biochemical- A gene that encodes a protein that can be extracted and observed; for example, isozymes and storage proteins.
- Cytological - The chromosomal banding produced by different stains; for example, G banding.
- DNA-based and/or molecular- A unique (DNA sequence), occurring in proximity to the gene or locus of interest, can be identified by a range of molecular techniques such as RFLPs, RAPDs, AFLP, DAF, SCARs, microsatellites etc.

Sax in 1923 first reported association of a simply inherited genetic marker with a quantitative trait in plants when he observed segregation of seed size associated with segregation for a seed coat color marker in beans (*Phaseolus vulgaris* L.). Rasmusson in 1935 demonstrated linkage of flowering time (a quantitative trait) in peas with a simply inherited gene for flower color.

Gene vs Marker

The gene of interest is directly related with production of protein(s) that produce certain phenotypes whereas markers should not influence the trait of interest but are genetically linked (and so go together during segregation of gametes due to the concomitant reduction in homologous recombination between the marker and gene of interest). In many traits genes are discovered and can be directly assayed for their presence with a high level of confidence. However, if a gene is not isolated marker's help is taken to tag a gene of interest. In such case there may be some false positive results due to recombination between marker of interest and gene (or QTL). A perfect marker would elicit no false positive results.

Important Properties of Ideal Markers for MAS

An ideal marker:

- Easy recognition of all possible phenotypes (homo- and heterozygotes) from all different alleles
- Demonstrates measurable differences in expression between trait types and/or gene of interest alleles, early in the development of the organism
- Has no effect on the trait of interest that varies depending on the allele at the marker loci
- Low or null interaction among the markers allowing the use of many at the same time in a segregating population

- Abundant in number
- Polymorphic.

Demerits of Morphological Markers

Morphological markers are associated with several general deficits that reduce their usefulness including:

- the delay of marker expression until late into the development of the organism
- dominance
- deleterious effects
- pleiotropy
- confounding effects of genes unrelated to the gene or trait of interest but which also affect the morphological marker (epistasis)
- rare polymorphism
- frequent confounding effects of environmental factors which affect the morphological characteristics of the organism.

To avoid problems specific to morphological markers, the DNA-based markers have been developed. They are highly polymorphic, simple inheritance (often codomimant), abundantly occur throughout the genome, easy and fast to detect, minimum pleiotropic effect and detection is not dependent on the developmental stage of the organism. Numerous markers have been mapped to different chromosomes in several crops including rice, wheat, maize, soybean and several others. Those markers have been used in diversity analysis, parentage detection, DNA fingerprinting, and prediction of hybrid performance. Molecular markers are useful in indirect selection processes, enabling manual selection of individuals for further propagation.

Selection for Major Genes Linked to Markers

The major genes which are responsible for economically important characteristics are frequent in the Plant Kingdom. Such characteristics include disease resistance, male sterility, self-incompatibility, others related to shape, color, and architecture of whole plants and are often of mono- or oligogenic in nature. The marker loci which are tightly linked to major genes can be used for selection and are sometimes more efficient than direct selection for the target gene. Such vantages in efficiency may be due for example, to higher expression of the marker mRNA in such cases that the marker is actually a gene.

Alternatively, in such cases that the target gene of interest differs between two alleles by a difficult-to-detect single nucleotide polymorphism, an external marker (be it another gene or a polymorphism that is easier to detect, such as a short tandem repeat) may present as the most realistic opti.

Situations that are Favorable for Molecular Marker Selection

There are several indications for the use of molecular markers in the selection of a genetic trait.

In such situations that:

- the selected character is expressed late in plant development, like fruit and flower features or adult characters with a juvenile period (so that it is not necessary to wait for the organism to become fully developed before arrangements can be made for propagation)

- the expression of the target gene is recessive (so that individuals which are heterozygous positive for the recessive allele can be crossed to produce some homozygous offspring with the desired trait)

- there is requirement for the presence of special conditions in order to invoke expression of the target gene(s), as in the case of breeding for disease and pest resistance (where inoculation with the disease or subjection to pests would otherwise be required). This advantage derives from the errors due to unreliable inoculation methods and the fact that field inoculation with the pathogen is not allowed in many areas for safety reasons. Moreover, problems in the recognition of the environmentally unstable genes can be eluded.

- the phenotype is affected by two or more unlinked genes (epistatis). For example, selection for multiple genes which provide resistance against diseases or insect pests for gene pyramiding.

The cost of genotyping (an example of a molecular marker assay) is reducing while the cost of phenotyping is increasing particularly in developed countries thus increasing the attractiveness of MAS as the development of the technology continues.

Steps for MAS

Generally the first step is to map the gene or quantitative trait locus (QTL) of interest first by using different techniques and then use this information for marker assisted selection. Generally, the markers to be used should be close to gene of interest (<5 recombination unit or cM) in order to ensure that only minor fraction of the selected individuals will be recombinants. Generally, not only a single marker but rather two markers are used in order to reduce the chances of an error due to homologous recombination. For example, if two flanking markers are used at same time with an interval between them of approximately 20cM, there is higher probability (99%) for recovery of the target gene.

QTL Mapping Techinques

In plants QTL mapping is generally achieved using bi-parental cross populations; a cross between two parents which have a contrasting phenotype for the trait of interest are developed. Commonly used populations are recombinant inbred lines (RILs), doubled haploids (DH), back cross and F_2. Linkage between the phenotype and markers which have already been mapped is tested in these populations in order to determine the position of the QTL. Such techniques are based on linkage and are therefore referred to as "linkage mapping".

Single Step MAS and QTL Mapping

In contrast to two-step QTL mapping and MAS, a single-step method for breeding typical plant populations has been developed. In such an approach, in the first few breeding cycles, markers linked to the trait of interest are identified by QTL mapping and later the same information in used in the same population. In this approach, pedigree structure are created from families that are created by crossing number of parents (in three-way or four way crosses).

Both phenotyping and genotyping is done using molecular markers mapped the possible location of QTL of interest. This will identify markers and their favorable alleles. Once these favorable marker alleles are identified, the frequency of such alleles will be increased and response to marker assisted selection is estimated. Marker allele(s) with desirable effect will be further used in next selection cycle or other experiments.

Paratransgenesis

Paratransgenesis is a technique that attempts to eliminate a pathogen from vector populations through transgenesis of a symbiont of the vector. The goal of this technique is to control vector-borne diseases. The first step is to identify proteins that prevent the vector species from transmitting the pathogen. The genes coding for these proteins are then introduced into the symbiont, so that they can be expressed in the vector. The final step in the strategy is to introduce these transgenic symbionts into vector populations in the wild.

The first example of this technique used *Rhodnius prolixus* which is associated with the symbiont *Rhodococcus rhodnii*. *Rhodnius prolixus* is an important insect vector of Chagas's disease that is caused by *Trypanosoma cruzi*. The strategy was to engineer *R. rhodnii* to express proteins such as Cecropin A that are toxic to *Trypanosoma cruzi* or that block the transmission of *Trypanosoma cruzi*.

5

Cells and Viruses—Overview

Cellular Organization

Cells are the smallest structural unit of living organisms, capable of maintaining life and reproducing. Viruses are not cells because they cannot maintain life and reproduce by themselves.

Although a nerve cell looks entirely different from a red blood cell, their organizations are essentially the same. Even plant cells and animal cells share significant similarity in the overall organization.

Classification of Cells and Organisms

All cells are divided into two types: prokaryotic cells and eukaryotic cells.

The prokaryotic cell does not have a nucleus.

The eukaryotic cell contains a nucleus.

Eukaryotes are the organisms made up of eukaryotic cells. They include protista, fungi, animals and plants. Prokaryotes include archaebacteria and eubacteria. They are single-cell organisms.

More recently, "archaebacteria" have been placed in a category outside "bacteria", because they are quite different from the ordinary bacteria. According to the new classification, prokaryotes are divided into archaea and bacteria, where "archaea" is equivalent to "archaebacteria", and "bacteria" is the same as "eubacteria". Archaea live in extreme environments. They may be organized into three groups:

> *Methanogens live in anaerobic environment such as swamps. They produce methane and cannot tolerate exposure to oxygen. Extreme halophiles live in very high*

concentrations of salt (NaCl), e.g., the Dead Sea and the Great Salt Lake. Extreme thermophiles live in hot, sulfur rich and low pH environment, such as hot springs, geysers and fumaroles in the Yellowstone National Park.

Basic Cellular Components

All cells contain cytoplasm, plasma membrane, and DNA.

Cytoplasm is the viscous contents of a cell, including proteins, ribosomes, metabolites and ions. Ribosomes are the sites of protein synthesis. Plasma membrane is the cell membrane surrounding cytoplasm. It consists of phospholipid bilayer, associated proteins and carbohydrates. Phospholipid bilayer is also the basic constituent of other biomembranes. DNA (deoxyribonucleic acid) is the genetic material. An eukaryotic cell contains several DNA molecules, located in the nucleus and mitochondria which are membrane-bound organelles. A prokaryotic cell contains a single DNA molecule, which has no specific boundary with the cytoplasm.

Prokaryotic Cells

A prokaryotic cell consists of DNA, cytoplasm, and a surface structure which includes the plasma membrane and some of the following components:

- Cell wall
- Capsule
- Slime
- Flagella
- Fimbriae/Pili.

In bacteria, the cell wall contains a unique structure called peptidoglycan. Archaea do not possess peptidoglycan, but some archaea may contain pseudopeptidoglycan, which is composed of N-acetyltalosaminuronic acid, instead of N-acetylmuramic in peptidoglycan. Because of this structural difference, archaea are resistant to many cell wall antibiotics.

Bacteria may be divided into two groups, on the basis of their cell wall structures and the response to Gram stain: Gram-negative and Gram-positive.

In Gram-negative bacteria, the cell wall is composed of three layers: (1) the periplasmic space which is an open area located outside the plasma membrane, (2) a thin layer of peptidoglycan external to the periplasmic space, and (3) the outer membrane surrounding the peptidoglycan. The Gram-positive bacteria do not have the periplasmic space and outer membrane, but have a thicker peptidoglycan layer. As a result, they are quite sensitive to lysozyme and penicillin. Capsule and slime are the hydrophilic gel surrounding the cell wall in most bacteria. The capsule is more closely associated with the cell than the slime. Flagella are long, rigid protein rods, facilitating the movement of motile bacteria. Fimbriae and pili are short hair-like structures used to attach other cells. They are essential for infecting other organisms.

Spores

The spore is a small, often unicellular, reproductive unit of plants, algae, fungi, protozoa, and bacteria. Bacterial spores have thick walls which can withstand varying temperatures, humidity, and other unfavorable conditions.

Eukaryotic Cells

The eukaryotic cell contains organelles, which are defined as membrane-bound structures such as nucleus, mitochondria, chloroplasts, endoplasmic reticulum (ER), Golgi apparatus, lysosomes, vacuoles, peroxisomes, etc. Prokaryotic cells do not have organelles. For animal cells, the cell surface consists of the plasma membrane only, but plant cells have an additional layer called cell wall, which is made up of cellulose and other polymers.

The nucleus is the largest organelle in an eukaryotic cell. It is not part of the cytoplasm. By definition, cytoplasm is everything inside the plasma membrane except the nucleus. Under microscope, the nucleus shows two distinct areas. The darker area is called nucleolus, and the lighter area is known as nucleoplasm. Cytosol is the cytoplasm excluding organelles. It contains cytoskeleton, ribosomes, proteins and other smaller molecules.

Biomembranes

All biological membranes, including plasma membranes and all organelle membranes, contain lipids and proteins. The lipids found in biomembranes are mainly phospholipids and cholesterol. In the plasma membrane and some of organelle membranes, proteins and phospholipids are attached to carbohydrates, forming glycoproteins and glycolipids, respectively.

Phospholipids

A phospholipid molecule consists of a hydrophilic polar head group and a hydrophobic tail. The polar head group contains one or more phosphate groups. The hydrophobic tail is made up of two fatty acyl chains. When many phospholipid molecules are placed in water, their hydrophilic heads tend to face water and the hydrophobic tails are forced to stick together, forming a bilayer.

Polar Head Groups

Most phospholipid head groups belong to phosphoglycerides, which contain glycerol joining the head and the tail. Examples of phosphoglycerides include phosphatidylcholine, phosphatidylserine, phosphatidylethanolamine, phosphatidylinositol, etc.

Fatty Acyl Chains

The fatty acyl chain in biomembranes usually contains even number of carbon atoms. They may be saturated (neighboring C atoms are all connected by single bonds) or unsaturated (some neighboring C atoms are connected by double bonds).

Table: Cellular fatty acids.

Chemical Formula	Name
Saturated fatty acid	
$CH_3(CH_2)_{10}COOH$	Lauric
$CH_3(CH_2)_{12}COOH$	Myristic
$CH_3(CH_2)_{14}COOH$	Palmitic
$CH_3(CH_2)_{16}COOH$	Stearic
$CH_3(CH_2)_{18}COOH$	Arachidic
$CH_3(CH_2)_{22}COOH$	Lignoceric
Unsaturated fatty acid	
$CH_3(CH_2)_5CH=CH(CH_2)_7COOH$	Palmitoleic
$CH_3(CH_2)_7CH=CH(CH_2)_7COOH$	Oleic
$CH_3(CH_2)_4CH=CHCH_2CH=CH(CH_2)_7COOH$	Linoleic
$CH_3(CH_2)_4(CH=CHCH_2)_3CH=CH(CH_2)_3COOH$	Arachidonic
$CH_3CH_2CH=CHCH_2CH=CHCH_2CH=CH(CH_2)_7COOH$	Linolenic

Note: In the bond-line representation,

Palmitic acid is represented as;

Arachidonic acid is represented as;

Cholesterol and Steroids

Cholesterol is absent from most prokaryotic cells, but abundant in the plasma membrane of mammalian cells. It is used as a precursor to generate other important steroids.

Cholesterol plays a central role in atherosclerosis - a disorder that may cause heart attack or stroke. It is also involved in Alzheimer's disease.

Glycoproteins and Glycolipids

Glycoproteins are the proteins covalently attached to carbohydrates such as glucose, galactose, lactose, fucose, sialic acid, N-acetylglucosamine, N-acetylgalactosamine, etc.

Glycolipids are carbohydrate-attached lipids. Their role is to provide energy and also serve as markers for cellular recognition.

The antigens which determine blood types belong to glycoproteins and glycolipids (more info).

Blood Group Antigens

The antigens which determine blood types belong to glycoproteins and glycolipids. There are three types of blood-group antigens: O, A, and B. They differ only slightly in the composition of carbohydrates.

The Nucleus

The cell nucleus consists of nuclear envelope, nucleolus and nucleoplasm. Most chromosomes are located in the nucleoplasm, but portions of several chromosomes containing clusters of rRNA genes may get together in the nucleolus, forming the nucleolar organizing region. The major role of the nucleolus is to produce rRNA.

Introduction to Chromosomes

Chromosomes are the structures that hold DNA molecules. One chromosome contains a DNA molecule. Each chromosome has a p and q arm; p is the shorter arm and q is the longer arm. The arms are separated by a pinched region called the centromere.

In order for chromosomes to be seen with a microscope, they need to be stained. Once stained, the chromosomes look like strings with light and dark "bands" and their picture can be taken. The picture, or chromosome map, is called a karyotype.

The germ cell (sperm or egg) of a human being contains 23 chromosomes, labeled from 1 to 22 and either X or Y. The somatic cells (cells other than germ cells) of a normal person has 46 chromosomes. For other species, the chromosome number varies from 1 to 1260.

A human somatic cell contains two chromosomes that determine the sex of a person. The two sex chromosomes are XY in male and XX in female. The gene (SRY) that is important for testis formation is located on chromosome Y. It is possible that a person with testis still exhibits female characteristics.

Chromosomes and Karyotype

In a non-dividing cell, chromosomes are not visible by light microscopy, because chromatin spreads throughout the nucleus. During the metaphase of cell division, the chromatin condenses and becomes visible as chromosomes. At this time, each chromosome has been duplicated. A chromosome becomes two sister chromatids attached at the centromere.

Chromosome Banding

To see chromosomes by microscope, they are normally treated with chemical dyes, such as Giemsa. The chromosome will appear as a series of alternate dark and light bands. If Giemsa is used, the dark band is called G-band or G-positive band, and the light band is named G-negative band. Similar banding patterns can be observed by using another dye, Quinacrine. However, if chromosomes were treated in a hot alkaline solution before staining with Giemsa, a reverse pattern will be observed, namely, the original dark band will become light band, and vice versa. For this reason, the G-negative band is also known as the R-band.

Chromosome bands are named as follows. Each chromosome consists of two arms separated by the centromere. The long arm and short arm are labeled q (for queue) and p (for petit), respectively. At the lowest resolution, only a few major bands can be distinguished, which are labeled q1, q2, q3; p1, p2, p3, etc., counting from the centromere. Higher resolution reveals sub-bands, labeled q11, q12, q13, etc. Sub-sub-bands identified by even higher resolution are labeled q11.1, q11.2, q11.3, etc. Traditionally, the short arm (p) is displayed on top of the long arm (q).

Karyotype

Karyotype is the representation of entire metaphase chromosomes in a cell, arranged in order of size.

Chromosome Numbers

A germ cell (sperm or egg) contains only one set of chromosomes. It belongs to haploid, represented as 1n. Somatic cells (cells other than germ cells) of sexually reproducing organisms are diploid, denoted by 2n. In humans, the haploid chromosome number is 23, but the diploid chromosome number is 46.

Extremes

Smallest number: The female of a subspecies of the ant, *Myrmecia pilosula*, has one pair of chromosomes per cell. Its male has only one chromosome in each cell.

Largest number: In the fern family of plants, the species *Ophioglossum reticulatum* has about 630 pairs of chromosomes, or 1260 chromosomes per cell.

Sex Chromosomes

Sex chromosomes determine the sex of an organism. A human somatic cell has two sex chromosomes: XY in male and XX in female. A human germ cell has one sex chromosome: X or Y in a sperm and X in an egg. When an X-sperm is combined with an egg, the resulting zygote (fertilized egg) will contain two X chromosomes. A person developed from the XX-zygote will have the characteristics of a female. Combination of a Y-sperm and an egg will produce a male.

The SRY Gene

Usually, a woman has two X chromosomes (XX) and a man one X and one Y (XY). However, both male and female characteristics can sometimes be found in one individual, and it is possible to have XY women and XX men. Analysis of such individuals has revealed some of the molecules involved in sex determination, including one called SRY, which is important for testis formation.

SRY (which stands for sex-determining region Y gene) is found on the Y chromosome. In the cell, it binds to other DNA and in doing so distorts it dramatically out of shape. This alters the proportion of the DNA and likely alters the expression of a number of genes, leading to testis formation. Most XX men who lack a Y chromosome do still have a copy of the SRY gene on one of their X chromosomes (moved there by chromosomal translocation). This copy accounts for their maleness. However, because the remainder of the Y chromosome is missing they frequently do not develop secondary sexual characteristics in the usual way.

Cytoplasmic Organelles

By definition, organelles are the membrane-bound structures in a cell. The nucleus is an example. Other organelles are located in the cytoplasm such as mitochondria, chloroplasts, endoplasmic reticulum, Golgi apparatus, peroxisomes, lysosomes, vacuoles and glyoxisomes.

Mitochondria

An eukaryotic cell contains many mitochondria, occupying up to a quarter of the cytoplasmic volume. The size of a mitochondrion is about 1.5-2 mm in length, 0.5-1 mm in diameter, approximately the same as *E. coli*. It has two membranes: outer membrane and inner membrane. Mitochondria also have their own DNA (represented as mtDNA), which encodes some of the proteins and RNAs in mitochondria. However, most proteins operating in mitochondria still originate from nuclear DNA.

The major role of mitochondria is to produce ATP (adenosine triphosphate), which carries high energy to power most cellular processes. Such energy is stored in the phosphoanhydride bonds of ATP. During ATP hydrolysis, the bond is broken, releasing 7.3 kcal/mole of energy. Many cellular processes can utilize the released energy by coupling with the ATP hydrolysis.

In animal cells, the major sources for the synthesis of ATP are fatty acids and glucose. Oxidation of an 18-carbon fatty acid can make 146 ATP molecules. By contrast, oxidation of one glucose molecule (6 carbons) can generate only 36 ATP molecules.

The generation of ATP involves a series of electron transport. Inevitably, electrons may leak from the electron transport chain, producing free radicals. This has been suggested to be the major mechanism involved in the aging process.

Chloroplasts

Like mitochondria, a chloroplast also contains both outer and inner membranes on its surface. Inside the chloroplast, there are many thylakoids, each is enclosed by a membrane. Chlorophylls are located on the thylakoid membrane to absorb light for photosynthesis.

In the first step of photosynthesis, light energy is used to split water into hydrogen ions and oxygen molecules. The generated hydrogen ions will create a concentration gradient across the thylakoid membrane. Movement of hydrogen ions through the membrane is coupled to ATP synthesis. The overall reactions can be written as

Endoplasmic Reticulum

Endoplasmic reticulum (ER) can be divided into rough ER and smooth ER. The major role of rough ER is to process the newly synthesized peptides from ribosomes. Therefore, the surface of rough ER is usually associated with ribosomes and thus appears "rough". Smooth ER is involved in the synthesis and metabolism of lipids. Hepatocytes are abundant in smooth ER.

Golgi Apparatus

Golgi apparatus is a major site for sorting and modifications of proteins and lipids. After proteins are sorted at rough ER, they are enclosed in transport vesicles and carried to the Golgi apparatus. Some proteins could be modified into glycoproteins and then transported to other destinations.

Peroxisomes

Peroxisomes contain enzymes for degrading amino acids and fatty acids. These reactions produce harmful hydrogen peroxide. Hence, peroxisomes also contain catalase to convert hydrogen peroxide into water and oxygen:

Lysosomes

The major function of lysosomes is to degrade various macromolecules in the cell. They contain nuclease for degrading DNA and RNA, protease for degrading proteins and other enzymes for degrading polysacchrides and lipids. Lysosomes exist only in animal cells. Although plant cells do not have lysosomes, their vacuoles are also capable of degrading macromolecules..

Vacuoles

Vacuoles store small molecules such as water, ions, sucrose and amino acids. They can also hold waste products which will be slowly degraded. They typically occupy more than 30% of the cell volume, but may expand up to 90%.

Glyoxisomes

Glyoxisomes are found mainly in plant seeds. Their major function is to convert fatty acids into acetyl CoA for the glyoxylate cycle where two acetyl-CoA molecules are converted to a 4-carbon dicarboxylic acid. Peroxisomes and glyoxisomes are also called microbodies.

Viruses

Viruses are the smallest organisms, with diameters ranging from 20 nm to 300 nm (1 nm = 10^{-9} meter). Viruses are not cells. They consist of one or more molecules of DNA or RNA, which contain the virus's genes, surrounded by a protein coat called capsid. Some viruses also have an envelope surrounding the capsid. Viruses can be sphere-shaped, or helical. Unlike most bacteria, most viruses do cause disease because they invade living, normal cells, such as those in the human body. They then multiply and produce other viruses like themselves. Each virus is very particular about which cell it attacks. Various human viruses specifically attack particular cells in the body's organs, systems, or tissues, such as the liver, respiratory system, or blood cells.

Although types of viruses behave differently, most survive by taking over the machinery that makes a cell work. Briefly, when a single virus particle, a "virion", comes in contact with a cell it likes, it may attach to special landing sites on the surface of that cell. From there, the virus may inject molecules into the cell, or the cell may swallow up the virion. Once inside the cell, viral molecules such as DNA or RNA direct the cell to make new virus offspring. That's how a virus "infects" a cell. Viruses can even "infect" bacteria. These viruses, called bacteriophages, may help researchers develop alternatives to antibiotic medicines for wiping out bacterial infections.

Many viral infections do not result in disease. For example, by the time most people in the United States become adults, they have been infected by cytomegalovirus (CMV). Most of these people, however, do not develop CMV disease symptoms. Other viral infections can result in deadly diseases, such as HIV which causes acquired immunodeficiency syndrome (AIDS) and coronaviruses which cause severe acute respiratory syndrome (SARS).

Baltimore Classification of Viruses

The Baltimore classification is based on genetic contents and replication strategies of viruses. The genetic material in all types of cells is double-stranded DNA, but some viruses use RNA or single-stranded DNA to carry genetic information.

According to Baltimore classification, viruses are divided into the following seven classes:

1. dsDNA viruses
2. ssDNA viruses
3. dsRNA viruses
4. (+)-sense ssRNA viruses

5. (-)-sense ssRNA viruses
6. RNA reverse transcribing viruses
7. DNA reverse transcribing viruses.

Where "ds" represents "double strand" and "ss" denotes "single strand".

The Life Cycle of Viruses

The life cycle of viruses may be divided into the following stages:

Attachment

Attachment is a specific binding between viral surface proteins and their receptors on the host cellular surface. This specificity determines the host range of a virus. For instance, the human immunodeficiency virus (HIV) attacks only human's immune cells (mainly T cells), because its surface protein, gp120, can interact with CD4 and chemokine receptors on the T cell's surface.

Penetration

Following attachment, viruses may enter the host cell through receptor mediated endocytosis or other mechanisms.

Uncoating

Uncoating is a process that viral capsid is degraded by viral enzymes or host enzymes.

Replication

Replication involves assembly of viral proteins and genetic materials produced in the host cell.

Release

Viruses may escape from the host cell by causing cell rupture (lysis). Enveloped viruses (e.g., HIV) typically "bud" from the host cell. During the budding process, a virus acquires the phospholipid envelope containing the embedded viral glycoproteins.

6

Enzymes in Molecular Biology Research

Since molecular cloning has become routine laboratory technique, manufacturers offer countless sources of enzymes to generate and manipulate nucleic acids. Thus, selecting the appropriate enzyme for a specific task may seem difficult to the novice. This review aims at providing the readers with some cues for understanding the function and specificities of the different sources of polymerases, ligases, nucleases, phosphatases, methylases, and topoisomerases used for molecular cloning.

We provide a description of the most commonly used enzymes of each group, and explain their properties and mechanism of action. By pointing out key requirements for each enzymatic activity and clarifying their limitations, we aim at guiding the reader in selecting appropriate enzymatic source and optimal experimental conditions for molecular cloning experiments.

Keywords: Enzymes, Molecular biology, Molecular cloning.

At a time when molecular cloning has become routine laboratory technique, we thought it was important to provide readers with some cues for understanding the function and specificities of the different enzymes used to generate and manipulate nucleic acids. Over the past few years, the tremendous expansion of cloning techniques and applications has triggered an enormous interest from laboratory suppliers. As a result, countless sources of enzymes are now available, and selecting the appropriate enzyme for a specific task may seem difficult to the novice.

Nucleic acids used for molecular cloning can be of natural or synthetic origin, and their length ranges from a few to several thousands nucleotides. Nucleic acids can be extensively manipulated, in order to acquire specific characteristics and properties. Such manipulations include propagation, ligation, digestion, or addition of modifying groups such as phosphate or methyl groups. These modifications are catalyzed by polymerases, ligases, nucleases, phosphatases, and methylases, respectively. In this review, we provide a description of the main enzymes of each group, and explain their properties and mechanism of action. Our goal is to give the reader a better understanding of the fundamental enzymatic activities that are used in molecular cloning.

DNA Polymerases (DNA-dependent DNA Polymerase, EC 2.7.7.7)

DNA polymerases are enzymes that catalyze the formation of polymers made by the assembly of multiple structural units or deoxyribo-nucleotides triphosphate (dNTPs). None of the DNA polymerases that have been characterized thus far can direct *de novo* synthesis of a polynucleotidic molecule from individual nucleotides. The DNA polymerases only add nucleotides to the 32 -OH end of a pre-existing primer containing a 52 -phosphate group. Primers are short stretches of RNA complementary to about 10 nucleotides of DNA at the 52 end of the molecule to replicate. Primers are synthesized by an RNA polymerase called primase.

Most polymerases used in molecular biology originate from bacteria or their infecting viruses (bacteriophages or phages). We will only discuss prokaryotic polymerases in this review. The functional and structural properties of eukaryotic DNA polymerases, which are specific for chromatin-embedded DNA, are reviewed elsewhere (Frouin et al. 2003; Garg and Burgers 2005).

Prokaryotic DNA Polymerases

In bacteria, three DNA polymerases act in concert to achieve DNA replication: Pol I, Pol II, and Pol III. All three enzymes catalyze 52 '!32 elongation of DNA strands in the presence of primer and dNTPs, but have variable elongation rate. DNA Pol III was discovered in 1970 (Kornberg and Gefter 1970) and is the main enzymatic complex driving prokaryotic replication. All three DNA Pol also have a 32 → 52 (reverse) exonuclease activity, otherwise known as proofreading activity as it initiates removal of incorrectly added bases as polymerization progresses. The proofreading activity increases fidelity but slows down polymerase progression. In addition to polymerase and proofreading activities common to all three DNA polymerases,

DNA Pol I also has a 52 '!32 exonuclease activity, which is used for the removal of RNA primer from the 52 end of DNA chains, and for excision-repair (upstream of polymerization). The 52 '!32 exonuclease activity of Pol I is utilized *in vitro* for nick-translation (*i.e.* tagging technique in which some of the nucleotides of a DNA sequence are replaced with labeled analogues).

Pol I and Klenow Fragment

The native DNA Pol I has been successfully used to remove 32 protruding DNA ends (in the absence of dNTPs), or to fill in cohesive ends (in the presence of dNTPs) before addition of molecular linkers. However, the 52 '!32 exonuclease activity of DNA Pol I makes it unsuitable for all applications that require polymerization activity alone (*e.g.* to fill in cohesive ends before addition of linkers, or to copy single-stranded DNA in the dideoxy method for sequencing). Fortunately, it was discovered that proteolytic digestion of *E. Coli* DNA Pol I (109 kDa) generates two fragments (76 and 36 kDa). The large fragment, also known as DNA Pol IK or Klenow fragment (named after its inventor, (Klenow and Henningsen 1970)), contains the 52 '!32 polymerase and 32 → 52 exonuclease (proofreading) activities of DNA Pol I, while the small fragment exhibits the 52 '!32 exonuclease activity alone. Since then, recombinant sources of Klenow significantly improved the functional quality of this fragment by eliminating contaminations due to the presence of residual native enzyme in proteolytically treated preparations.

T4 DNA Polymerase

Bacteriophage T4 polymerase requires a template and a primer to exhibits two activities: it is a 32 '!52 (reverse) exonuclease in the absence of dNTPs, and a 52 '!32 polymerase in the presence of dNTPs. Unlike the *E. Coli* DNA Pol I, T4 DNA Pol does not exhibit a 52 '!32 exonuclease activity (Englund 1971). Therefore, T4 DNA Pol can be used instead of Klenow to fill in 52 -protruding ends of DNA fragments, for nick translation, or for labelling 32 ends of duplex (double-stranded) DNA. T4 DNA Pol exonuclease rate is approximately 40 bases per minutes on double stranded DNA, and about 4,000 bases on single stranded DNA. T4 DNA Pol polymerization rate reaches 15,000 nucleotides per minutes when assayed under standardized conditions.

Modified Bacteriophage T7 DNA Polymerase

A chemically modified phage T7 DNA polymerase has been described by Richardson et al. as an ideal tool for DNA sequencing

(Huber et al. 1987; Tabor et al. 1987; Tabor and Richardson 1987). Modified bacteriophage T7 DNA Pol is a complex of two proteins: the 84 kDa product of the T7 gene 5, and the 12 kDa *E. Coli* thioredoxin (Mark and Richardson 1976; Modrich and Richardson 1975).

The T7 gene 5 protein provides catalytic properties to the complex, while the thioredoxin protein connects the T7 gene 5 protein to a primer template, which allows the polymerization of thousands of nucleotides without dissociation, thereby increasing the efficiency of the T7 polymerase. Hence, modified T7 DNA Pol has a polymerization rate of more than 300 nucleotides per second, which makes it more than 70 times faster than that of AMV reverse transcriptase. Thus, this enzymatic complex can be used for preparation of radioactive probes and amplification of large DNA fragments.

Further characterization identified a 28-amino acid region (residues 118–145) as essential to T7 DNA Pol 32 '!52 exonuclease activity (Tabor and Richardson 1989). *In vitro* mutagenesis of the corresponding nucleotides in the T7 gene 5 resulted in complete elimination of the exonuclease activity, thereby increasing polymerase efficiency (9-fold) and spontaneous mutation rate (14-fold) (Tabor and Richardson 1989).

The mutant T7 polymerase/thioredoxin complex, commercially available under the name sequenase (United States Biochemical Corporation), is used for DNA sequencing because of its high efficiency and ability to incorporate nucleotide analogs (such as 52 -(α-thio)-dNTPs, dc7-GTP, or dITP used to improve the resolution of DNA sequencing gels, and to avoid gel compression resulting from base pairing).

Terminal Deoxynucleotidyl Transferase (DNA Nucleotidyl-exotransferase, EC 2.7.7.31)

Terminal deoxynucleotidyl transferase (TdT), initially purified from calf thymus (Krakow et al. 1962), is a DNA polymerase that catalyzes the addition of a homopolymer tail to 32 -OH ends of DNA, in a template-independent manner. TdT is used in molecular biology for labelling DNA 32 ends with modified nucleotides (such as ddNTP, DIG-dUTP, or radiolabeled nucleotides), for primer extension, or for DNA sequencing. It is also used in TUNEL (TdT dUTP Nick End Labelling) assay for the demonstration of apoptosis (Gavrieli et al. 1992).

TdT requires a single-stranded DNA primer in the presence of Mg^{2+} (three nucleotide-long minimum), but can accept double-stranded DNA as a primer in the presence of cobalt ions (Roychoudhury et al. 1976). However, addition of Co^{2+} may result in a relaxation of the helical structure of the DNA, thereby allowing the tailing of internal nicks. Use of Mg^{2+} reduces this problem but results in a significant decrease of the tail length (approximately one-fifth of the length obtained in the presence of Co^{2+} at identical enzyme:DNA ratios). Hence, optimization of incubation conditions is critical to control specificity, reactivity, and activity of TdT.

Thermostable DNA Polymerases

Although thermostable DNA polymerases were purified in the early seventies, their considerable interest for molecular cloning emerged from the development of polymerase chain reaction (PCR) and subsequent need for enzymes able to perform DNA synthesis at high temperature. Since thermostable polymerases are functional at high temperature, they can replicate DNA regions with high G/C content, similar to those frequently found in thermophilic organisms (high C/G content sequences form secondary structures that need to be properly denatured in order to be efficiently copied).

Bst Polymerase

Bst polymerase is a thermostable DNA Pol I that was isolated in 1968 (Stenesh et al. 1968) from the thermophilic bacterium *Bacillus stearothermophilus* (Bst), which proliferates between 39 and 70°C. Bst Pol I is active at an optimal temperature of 65°C, and is inactivated after 15 min incubation at 75°C. Bst Pol I possesses a 52 '!32 exonuclease activity and requires high Mg^{2+} concentration for maximum activity.

Protease digestion of Bst DNA Pol I generates two protein fragments. The large protein fragment of Bst DNA Pol I is thermostable, and thus very useful for sequencing reactions performed at 65°C to avoid problems due to hairpin formation. Like Klenow, the Large Fragment of Bst DNA Pol I shows a faster strand displacement than its full length counterpart. Recombinant Bst DNA Pol I is presently available from Epicentre Biotechnologies, which also commercializes the Large Fragment of rBst DNA Pol I under the name IsoTherm™.

Taq Polymerase

Taq polymerase was first purified in 1976 (Chien et al. 1976) from a bacterium discovered 8 years earlier in the Great Fountain region of Yellowstone National Park. This bacterium, which thrives at 70°C and survives at temperatures as high as 80°C, was named *Thermophilus aquaticus* (*T. aquaticus* or Taq). Taq polymerase has a halflife of 40 min at 95°C, and 5 min at 100°C, which allows the PCR reaction. For optimal activity, Taq polymerase requires Mg^{2+} and a temperature of 80°C. Taq polymerase lacks 32 '!52 exonuclease (proofreading) activity, and is therefore often described as being a low fidelity enzyme (error rate between 1×10^{-4} and 2×10^{-5} errors per base pair, depending on experimental conditions). Yet, one should keep in mind that this corresponds to a quite good accuracy (inverse of the error rate) since 45,000 nucleotides can be incorporated into newly synthesized DNA strands before an error occurs.

Like other DNA polymerases lacking 32 '!52 exonuclease activity, Taq polymerase exhibits a deoxynucleotidyl transferase activity that is accountable for the addition of a few adenine residues at the 32 -end of PCR products. Later on, additional thermostable DNA polymerases have been discovered and commercialized. These enzymes have better accuracy than the original Taq. However, the term "Taq" is commonly used in place of "thermostable DNA polymerase", hence the erroneous term "high fidelity Taq", often used in laboratories.

Tth Polymerase

Tth polymerase, isolated from *Thermus thermophilus HB-8* (Ruttimann et al. 1985), is 94 kDa thermostable polymerase lacking 32 '!52 exonuclease (proofreading) activity. Tth polymerase catalyzes the polymerization of nucleotides into duplex DNA from a DNA template (DNA polymerase) in the presence of Mg^{2+}, and from an RNA template (reverse transcriptase) in the presence of Mn^{2+}.

Thermostable DNA Polymerases with Proofreading Activity

Several DNA polymerases isolated from thermophillic organisms exhibit 32 '!52 exonuclease (proofreading) activity, which increases fidelity. Among them, DNA polymerase from the archebacteirum *Pyrococcus furiosus* (Pfu) has an error rate (1.6×10^{-6}) 10-fold lower than that of Taq polymerase. It can be purchased from numerous providers. Pow polymerase (isolated from *Pyrococcus woesei*) has a half-life greater than 2 h at 100°C and a very low error rate of 7.4×10^{-7}. Pow polymerase provides high PCR yields only for templates shorter than 3.5 kb. To circumvent PCR product size limitation, a mixture of Taq and Pow DNA polymerases has been commercialized by Roche Molecular Chemicals (under the name "Expand high fidelity"). This system allows amplification and cloning of long stretches of DNA (20–35 kb), and represents a unique tool for analyzing and sequencing large eucaryotic genes, or introducing large DNA fragments in lambda phages or cosmid vectors. Vent polymerase (also known as Tli polymerase since isolated from *Thermococcus litoralis*) has an error rate comprised between that of Taq and Pfu.

Vent polymerase has a half-life of 7 h at 95°C. It is marketed by New England Biolabs, together with a modified version that lacks exonuclease (proofreading) activity. Two DNA polymerase (Pol I and II) that exhibit a 32 '!52 exonuclease proofreading activity were also isolated from *Pyrococcus abyssi* (Gueguen et al. 2001). Pol I DNA polymerase from *P. abyssi* (Pab), marketed by Qbiogen as "Isis DNA Polymerase", has an error rate (0.66×10^{-6}) similar to that of Pfu. However, Pab is more thermostable than Pfu or Taq (Pab has a half-life of 5 h at 100°C), therefore being very useful for conducting PCR reactions that involve high temperature incubations.

Most of DNA polymerases and reverse transcriptases (described below) are commercially available in "Hot Start" (or equivalent) version. Hot Start modification is aimed at preventing non-specific priming events such as template/primer hybridization or primer dimmer formation, which occur in low stringency conditions (during PCR preparation). Non-specific priming events generate secondary products during the preparation and in the first cycle of the PCR, and are further amplified in subsequent cycles.

Hot Start technology introduces a "barrier" between secondary structures and the polymerase DNA-binding site. The "barrier", which can be an antibody, an oligonucleotide, or a reversible chemical modification of the polymerase's amino acids, is released from the enzyme during the first denaturation cycle of the PCR, thus restoring enzymatic activity only after denaturation of secondary structures. Hot Start modifications are very efficient in limiting polymerase activity at room temperature, and thereby facilitate PCR preparation while enhancing PCR specificity.

Reverse Transcriptase (RNA-dependent DNA Polymerase EC 2.7.7.49)

Until the 1960's, the transfer of genetic information was thought to flow unidirectionally from DNA to RNA (Crick 1958). Characterization of reverse transcriptase (RT) from Rous Sarcoma Virus (RSV) (Temin and Mizutani 1970) and Rauscher Leukaemia virus (RLV) (Baltimore 1970) exemplified that single stranded DNA could be synthesized from an RNA template. The resulting single-stranded DNA is a "complementary" copy of the RNA template, or cDNA.

Viral Reverse-transcriptases

RSV and RLV belong to the group of retroviridae (retroviruses). The life cycle of retroviruses includes an RT-directed transcription of their RNA genome and formation of a proviral double stranded molecule of DNA, which is integrated in the host genome. Infectious viral progeny is produced from transcription of the proviral DNA.

Many viruses contain a reverse transcription stage in their replication cycle.

Metaviridae are closely related to retroviruses and exist as retrotransposons in the eucaryotic host genome. Retrotransposons are mobile elements that amplify (multiply) through intermediate RNA molecules, which are reverse transcribed and integrated at new places in the host genome. Examples of metaviridae include the Saccharomyces cerevisiae Ty3 virus, Drosophila melanogaster gypsy virus, and Ascaris lumbricoides Tas virus.

Pseudoviridae (*e.g.* Saccharomyces cerevisiae Ty1 virus and Drosophila melanogaster copia virus) have a segmented single stranded RNA genome.

Hepadnaviridae (like the hepatitis B virus (Seeger and Mason 2000)) have a genome made of two uneven strands of DNA that exist as stretches of both single and double stranded circular DNA. The viral polymerase, which catalyses RNA- and DNA-dependent DNA synthesis, possesses both RNase H and protein priming activities. Upon infection, the relaxed circular DNA is converted into a circular DNA that is transcribed by the host RNA polymerase. This "pre-genomic" RNA is retrotranscribed in cDNA by the viral RT to give rise to the genomic viral DNA strands.

Caulimoviridae are unenveloped viruses that infect plants, including cauliflower (cauliflower mosaic virus) and soybean (soybean chlorotic mottle virus). Upon infection, the polyadenylated viral RNA is reverse-transcribed by the viral RT to give rise to genomic double-stranded DNA molecules. Viral RNA is also used for viral protein synthesis in the cytoplasm of infected cells. Although caulimoviridae's life cycle resembles that of retrovirus, replication of caulimoviridae does not involve the integration of proviral DNA in the host genome.

Non-viral Reverse Transcriptases

Transposons Nearly all organisms contain variable amounts of repetitive mobile DNA known as transposable elements (TEs) or transposons. TEs constitute more than 80% of the total genome in plants, while it represents about 45% of the human genome. Finnegan's classification (Finnegan 1989) distinguishes RNA transposons (Class I or retrotransposons), which amplify through a "copy and paste" type of transposition, and DNA transposons (Class II), which use a "cut and paste" type of transposition. However, the newer Wicker's classification takes into account the recent discovery of bacterial and eukaryotic TEs that copy and paste without RNA intermediates, and of new miniature inverted repeat transposable elements (MITEs) (Wicker et al. 2007).Retrotransposons, which make up to 42% of human genome, encode an RT used to make a cDNA copy of an RNA intermediate, which is produced after transcription of the retrotransposon integrated in the genome.

Telomerase Telomeres are non-coding, linear sequences that cap the ends of DNA molecules in eukaryotic chromosomes. Telomeres are made of up to 2,000 repeats of TTAGGG stretches. In normal cells, the DNA replication machinery is unable to duplicate the complete telomeric DNA. As a result, telomeres are shortened after every cell division. Having reached a critical length, telomeres are recognized as double strand break DNA lesions, and cells eventually enter senescence.In embryonic and adult stem cells, which have an extended lifespan, telomere length is maintained through activation of telomerase. Telomerase is a ribonucleoproteic complex that contains an RNA (TElomere RNA Component, or TERC) and an RT (TElomere Reverse Transcriptase, or TERT). The RNA provides the AAUCCC template directing the synthesis of TTAGGG repeats by the RT. Interestingly, telomerase is also activated during carcinogenesis (Raynaud et al. 2008).

Reverse Transcriptase for Molecular Biology

The use of RT in fundamental and applied molecular biology has been propelled by the introduction of the Reverse Transcription Polymerase Chain Reaction (RT-PCR). As a result, commercial sources of RT have flourished over the past two decades.

Because RTs are deprived of 32 '!52 exonuclease (proofreading) activity, they have much lower fidelity than DNA-dependent DNA polymerases. For instance, HIV RT has an error rate of 1 mutation per 1,500 nucleotides, whereas RT of avian and murine origin generate 1 mutation per 17,000 and 30,000 nucleotides, respectively.

AMV/MAV RT

The RT the most commonly used in molecular biology is the one that allows the replication of the Avian Myeoloblastosis Virus (AMV), an alpha retrovirus that induces myeoloblastosis in chicken (Baluda et al. 1983). Interestingly, cloning of the AMV genome identified the v-myb oncogene as responsible for intense proliferation of transformed myeloblasts. Insertion of v-myb oncogenic sequences in the AMV genome interrupts the coding sequence of the RT, thus making AMV RT-deficient. Like every deficient viruses, AMV depends on a helper virus for its replication. AMV helper, the Myeoloblastosis Associated Virus (MAV), is in fact the "real" source of RT in the life cycle of AMV, and for molecular biology (Perbal 2008).

The AMV/MAV RT is composed of two structurally related sub-units designated α and β (65 kDa and 95 kDa, respectively). The α subunit of the enzyme provides RT and RNase H activities. The RT activity requires the presence of a primer and a template. AMV/MAV RT is widely used to copy total messenger RNAs using polydT or random primers. RNase H activity is generated by proteolytic cleavage of the α subunit and associated with a 24 kDa fragment. RNase H is a processive exoribonuclease that degrades specifically RNA strands in RNA— DNA hybrids in either 52 '!32 or 32 '!52 directions.

The use of reverse transcriptase has found many applications in molecular cloning, two well-documented examples being the synthesis of cDNA from RNA in the preparation of expression libraries, as a first step for quantitative PCR, and for nucleotide sequencing. Several protocols have establish optimal conditions for high yields reactions (Berger et al. 1983), or for the synthesis of large RNA templates (Retzel et al. 1980).

MuLV RT

The *pol* gene of Moloney murine leukemia virus (M-MuLV) encodes an RT lacking DNA endonuclease activity, and exhibiting a lower RNAse H activity than AMV/MAV RT (Moelling 1974). M-MuLV RT is 4-times less efficient, and at least 4-times less stable than AMV/MAV RT. Therefore, comparable yields of cDNA synthesis require six to eight times more M-MuLV RT than AMV/MAV RT. However, M-MuLV RT is able to generate longer transcripts than does AMV/MAV RT when used in excess (Houts et al. 1979).

Thermostable Reverse Transcriptases

Several RTs identified in thermophilic organisms are commercially available and are useful in certain challenging conditions (*e.g.* to transcribe templates with high CG contents or abundant secondary structures).

Thermus thermophilus (Tth) is a DNA polymerase that efficiently reverse-transcribes RNA in the presence of $MnCl_2$. Upon chelation of the Mn^{2+} ions, its DNA polymerase activity allows for PCR amplification in the presence of $MgCl_2$.

Epicentre's MonsterScript™ RT is an Mg^{2+}-dependent thermostable RT that lacks RNase H activity, and is fully active at temperatures up to 65°C. According to its manufacturer, this enzyme can produce cDNA larger than 15 kb.

The Klenow fragment of *Carboxydothermus hydrogenoformans* (*C. therm.*) polymerase is a Mg^{2+}-dependent RT that is active at temperatures up to 72°C. It is marketed by Roche (C. therm. Mix) for RT PCR uses.

Thermo-X™ RT from Invitrogen has a half-life of 120 min at 65°C, which is the highest stability reported so far.

RNA Polymerase (DNA Dependent-RNA Polymerases)

DNA-dependent RNA polymerases catalyze the 52 '!32 elongation of RNA copies from DNA templates, a process called transcription. Like DNA polymerases that lack proofreading exonuclease activity, RNA polymerases can add an extra base at the end of a transcript.

The two main RNA polymerases used in molecular biology are SP6- and T7- RNA polymerases. SP6 RNA polymerase is a 96 kDa polypeptide purified from SP6 bacteriophage-infected *Salmonella typhimurium* LT2, while T7 RNA polymerase is a 98 kDa polypeptide (Stahl and Zinn 1981) produced by the T7 bacteriophage. RNA polymerases are used for *in vitro* synthesis of anti-sense RNA transcripts (Melton et al. 1984), production of labeled RNA probes, or for RNase protection mapping (Zinn et al. 1983).

SP6 and T7 RNA polymerases are similar enzymes: both require Mg^{2+} and a double-stranded DNA template, and are greatly stimulated by spermidine and serum albumine (Butler and Chamberlin 1982). In addition, T7 RNA polymerase typically requires the presence of dithiothreitol in reaction buffer.

SP6 and T7 RNA polymerases are extremely promoter specific and will transcribe any DNA sequence cloned downstream of their respective promoter (SP6 and T7). Importantly, SP6- or T7- RNA polymerase transcription proceeds through poly(A) stretches (Melton et al. 1984), and thus can progress around a circular template multiple times before disassociating. Therefore, linearization of the template prior to translation (a blunt or 52 overhang is recommended (Schenborn and Mierendorf 1985)) will guaranty an efficient termination of transcription. The RNA produced under these conditions is biologically active (Krieg and Melton 1984) and can be properly spliced (Green et al. 1983).

Ligases

DNA Ligases (EC 6.5.1.1 and 6.5.1.2 for ATP and NAD⁺ DNA Ligases, Repectively): DNA ligases connect DNA fragments by catalyzing the formation of a phosphodiester bond between a 32 -OH and a 52 -phosphate group at a single-strand break in double-stranded DNA (Lehman 1974). In cells, DNA ligases are essential for joining Okazaki fragments during replication, and in the last step of DNA repair process. DNA ligases are used in molecular biology to join DNA fragments with blunt or sticky ends such as those generated by restriction enzyme digestion, add linkers or adaptors to DNA, or repair nicks.

DNA ligases operate in a three-step reaction. The fist step involves the creation of a ligase-adenylate intermediate, in which a phosphoamide bond is created between a lysine residue and one AMP molecule of the enzyme cofactor (ATP or NAD⁺). Second, the AMP is transferred to the 52 -phosphate end of the DNA nick to form a DNA-adenylate (AppDNA). Finally, a nucleophilic attack from the 32 end of the DNA nick directed to the AppDNA results in joining of the two polynucleotides and release of AMP.

Original observations suggested that bacterial DNA ligases use NAD⁺ as a cofactor whereas DNA ligases from eukaryotes, viruses and bacteriophages use ATP. However, it is known now that some ligases can accept either cofactor, even though both cofactors are not equally efficient.

Non-thermostable DNA Ligases

The smallest known DNA ligases are the ATP-dependant DNA ligase from Chlorella virus and bacteriophage T7 (34 and 41 kDa, respectively). They are much smaller than eukaryotic DNA ligases, which can reach 100 kDa in size. While T7 and chlorella-encoded ATP-dependent ligases both contain only a nucleotidyl-transferase domain and an OB-fold domain, the eukaryotic ligases contain additional domains that include zinc fingers and BRCT (C-terminal portion of BRCA-1) domains with nuclear or mitochondrial localization signals.

The ligase the most frequently used in molecular biology is the bacteriophage T4 DNA ligase. T4 DNA ligase is a 68 kDa monomer that requires Mg^{2+} and ATP as cofactors. T4 DNA ligase can connect blunt and cohesive ends, or repair single stranded nicks in duplex DNA, RNA, or DNA/RNA hybrids.

The *E. Coli* DNA ligase works preferentially on cohesive double-stranded DNA ends. However, it is also active on blunt ends DNA in the presence of Ficoll or polyethylene glycol. Hybrids such as DNA-RNA or RNA-RNA are not efficiently formed by *E. Coli* DNA ligase. This can be used as an advantage when double stranded DNA ligation is wanted and blunt end ligation needs to be avoided.

Because DNA ligases' activity depend on several factors (such as temperature, fragment concentration, nature of fragments — blunt or sticky ends, length of sticky end, stability of hydrogen bonded structure —), it is difficult to presume the ideal incubation conditions for a specific ligation. For instance, ligation of fragments generated by Hind III is 10–40 times faster than the one for fragments generated with Sal I (even though both enzymes generate sticky fragments). For these reasons, the definition of the ligase unit is very specific: one ligation unit is defined by most suppliers as the amount of enzyme that catalyzes 50% ligation of Hind III fragments of lambda DNA in 30 min at 16°C under standard conditions.

Most standard ligation protocols recommend incubating the ligation reaction overnight at 16°C. However, ligation protocols should be empirically optimized for every ligation and according to the amount of DNA present in the reaction (although most suppliers give guidelines according to reaction volume, not DNA concentration). Typically, successful ligation with T4 DNA ligase have been reported with incubation varying from 10 min at room temperature to 24 h at 4°C. In addition, it is good practice to verify the activity of a ligase preparation being kept at "20°C for a long period of time, performing a periodical ligation test. A typical test for ligation of sticky ends is performed with Hind III-digested ë DNA. For blunt end ligations, the same procedure can be used with Hae III-digested DNA.

Thermostable DNA Ligases

Thermostable DNA ligases can perform ligation of duplex molecules and repair of single stranded nicks at temperatures ranging from 45 to 80°C. They are highly specific and are very well suited for applications that need high stringency ligations. Thermostable DNA ligases are isolated from diverse sources such as *Thermus thermophilus, Bacillus stearothermophilus* (Brannigan et al. 1999), *Thermus scotoductus* (Jonsson et al. 1994), and *Rhodothermus marinus*.

Thermostable DNA ligases are usually not a substitute for T4 or *E. Coli* DNA ligases but are used for very specific techniques such as Ligase Chain Reaction (LCR). LCR is a technique used to detect single base mutations: a primer is synthesized in two fragments that cover both sides of a possible mutation. Thermostable ligase will connect the two fragments only if they match exactly to the template sequence. Subsequent PCR reactions will amplify only if the primer is ligated.

RNA Ligases (EC 6.5.1.3)

RNA ligases catalyze the ATP-dependent formation of phosphodiester bonds between 52 -phosphate and 32 -OH termini of single stranded RNA or DNA molecules. In cells, they act mainly to reseal broken RNAs. Like DNA ligases, RNA ligases operate in a three-step reaction. First, the RNA ligase reacts with ATP to form a covalent ligase-(lysyl-N)-AMP intermediate, and pyrophosphate. Then the AMP moiety of the ligase adenylate is transferred to the 52 -phosphate end of the RNA to form an RNA-adenylate intermediate (AppRNA). In the third step, a nucleophilic attack of the 32 -OH end of the RNA on the AppRNA creates a phosphodiester bond, which seals the two RNA ends.

T4 RNA Ligases

The best characterized RNA ligase is the T4-bacteriophage RNA ligase (gp63), which was identified in T4-infected *E. Coli*. T4 RNA ligase is also the most commonly used RNA ligase in molecular biology. T4 RNA ligase 1 is used to ligate single stranded nucleic acids and polynucleotides to RNA molecules, usually to label RNA molecules at the 32 -end for RNA structure analysis, protein binding site mapping, rapid amplification of cDNA ends (RLM-RACE), ligation of oligonucleotide adaptors to cDNA, oligonucleotide synthesis (Kaluz et al. 1995), 52 nucleotide modifications of nucleic acids, and for primer extension for PCR. T4 RNA ligase can also be used to circularize RNA and DNA molecules. This enzyme is used in Ambion's FirstChoice™ RLM-RACE Kit for tagging the 52 ends of mRNA with oligonucleotide adaptor. A second RNA ligase encoded by the bacteriophage T4 has recently been described (Ho and Shuman 2002). T4 RNA ligase 2, also known as T4 Rnl-2 (gp24.1), catalyzes both intramolecular and intermolecular RNA strand ligation. Unlike T4 RNA ligase 1, T4 RNA ligase 2 is much more active joining nicks on double stranded RNA than on joining the ends of single stranded RNA. T4 RNA ligase 2 ligates 32 -OH/52 -phosphate RNA nicks, and can also ligate 32 -OH of RNA to the 52 -phosphate of DNA in a double stranded structure.

A truncated form of T4 RNA ligase 2 (truncated T4 RNL2, also known as RNL2 [1–249]) is commercialized by New England Biolabs. Truncated T4 RNL2, first 249 amino acids of the full length T4 RNA Ligase 2 (Ho and Shuman 2002), is unable to perform the first adenylation step of the ligation reaction. Thus, the enzyme does not require ATP but does need the pre-adenylated substrate to specifically ligates pre-adenylated 52 end of DNA or RNA to 32 end of RNA molecules. Truncated T4 RNL2 reduces background ligation because it selects adenylated primers. This enzyme can be use for optimized linker ligation for the cloning of microRNAs. An RNA ligase coding frame has also been identified in the pnk/pnl gene (ORF 86) from the baculovirus Autographa californica nucleopolyhedrovirus (ACNV) and from the radiation-resistant bacterium Deinococcus *radiodurans* (Dra) (Martins and Shuman 2004b). DraRnl ligates 32 -OH/52 -phosphate RNA nicks that can occur in either duplex RNA or in RNA/DNA hybrids. However, it cannot ligate nicks in DNA molecules, as it requires a ribonucleotidic 32 -OH.

Thermostable RNA Ligases

Thermostable RNA ligases have been isolated from bacteriophages rm378 and *TS2126* that infect the eubactrium *rhodothermus marinus* and bacterium *thermus scotoductus* respectively.

Phosphate Transfer and Removal

The principal characteristics of the enzymes used for phosphate transfer and removal in molecular biology discussed in this section.

Alkaline Phosphatase (EC 3.1.3.1)

Alkaline phosphatase is purified from either *E. Coli* or higher organisms (*e.g.* calf intestine). It is used for removal of 52 -phosphate groups from nucleic acids in order to prevent recircularization of DNA vectors in cloning experiments. Alkaline phosphatase does not hydrolyze phosphodiester bonds.

Although *E. Coli* and calf intestine alkaline phosphatases have different structures (80 and 140 kDa, respectively), both enzymes are zinc and magnesium containing protein (Reid and Wilson 1971). They are inactivated by chelating agents such as EGTA, and by low concentrations of inorganic phosphate. Importantly, inorganic phosphates need to be removed after restriction endonuclease cleavage and prior to incubation with alkaline phosphatase. Inorganic phosphates are certainly removed if cloning strategy involves gel migration and purification of linearized vector. If no purification is made, dialysis of endonuclease-digested DNA is required prior to alkaline phosphatase treatment (ethanol precipitation does not efficiently remove inorganic phosphates). Similarly, when labelling DNA fragments at the 52 -end (*e.g.*[32]P labelling), incubation with alkaline phosphatase needs to precede incubation with polynucleotide kinase in the presence of [^{32}P]deoxynucleoside triphosphate.

T4 Polynucleotide Kinase (Polynucleotide 52 -hydroxy Kinase, EC 2.7.1.78)

Polynucleotide kinase catalyzes the transfer of a phosphate group from an ATP molecule to the 52 -OH terminus of a nucleic acid (DNA or RNA, with no size limitations), the exchange of 52 -phosphate groups, or the phosphorylation of 32 -ends of mononucleotides. Encoded by the *pse* T gene of bacteriophage T4 (Depew et al. 1975), the T4 polynucleotide kinase is a tetramer of four identical 33-kDa monomers. Polynucleotide kinase is commonly used for labelling experiments with radiolabeled ATP utilized as a phosphate donor.

Phosphate transfer activity is optimum at 37°C, pH 7.6, in the presence of Mg^{2+} and reducing reagents (DTT or β2-mercaptoethanol), and with a minimum of 1 mM ATP and a 5:1 ratio of ATP over 52 -OH ends (Lillehaug et al. 1976). When substrate concentration is limited, addition of 6% polyethylene glycol (PEG 8000) in the reaction mixture enhances radiolabeling of recessed, protruding, and blunt 52 -termini of DNA (Harrison and Zimmerman 1986).

When radiolabeled ATP is used (*e.g.* ã[32]P-ATP), polynucleotide kinase generates labeled DNA or RNA molecules with ^{32}P at their 52 ends, by catalyzing either direct phosphorylation of 52 -OH groups generated after alkaline phosphatase digestion, or exchange of the 52

-phosphate groups. This reaction is efficiently used for labelling DNA or RNA strands prior to base specific sequencing. Polynucleotide kinase is also used for mapping of restriction sites, DNA and RNA fingerprinting, and synthesis of substrates for DNA or RNA ligase. In addition to the kinase activity described above, T4 polynucleotide kinase also exhibits a 32 -phosphatase activity (Cameron and Uhlenbeck 1977). Optimum pH for 32 phosphatase activity is comprised between 5 and 6, *i.e.* more acidic than for phosphate transfer activity. Based on this property, protocols using T4 polynucleotide kinase as specific 32 -phosphatase have been developed (Cameron et al. 1978).

A T4 polynucleotide kinase lacking 32 -phosphatase activity has been purified from *E. Coli* infected with a mutant T4 phage producing an altered *pse* T1 gene product (Cameron et al. 1978). Like the wild-type enzyme, the mutant polynucleotide kinase is made of four subunits of 33 kDa each, effectively transfers the gamma phosphate of ATP to the 52 -OH terminus of DNA and RNA. In addition, the mutant and the wild type polynucleotide kinases require similar magnesium ion concentrations, have the same pH optima and are both inhibited by inorganic phosphate. The mutant polynucleotide kinase is very useful when the 32 -exonuclease activity must be avoided (*e.g.* 52 -^{32}P terminal labelling of 32 -CMP in view of 32 end-labelling of RNA species prior to fingerprinting or sequencing).

Tobacco Acid Pyrophosphatase

Tobacco acid pyrophosphatase is used as a first step for labelling mRNAs at their of 52 ends, which are usually capped, in order to generate radiolabeled probes of for RNA sequencing: *in vitro*, ^{32}P-labeling of mRNA 52 terminus requires the elimination of the 7-methylguanosine and 52 -phosphate moieties of the capped end. Tobacco acid pyrophosphatase hydrolyzes the pyrophosphate bond in the cap's triphosphate bridge, generating a 52 -phosphate terminus on the RNA molecule (leading to p7MeG, pp7MeG, and ppN— pN— mRNA). The generated open cap can then be dephosphorylated by alkaline phosphatase, and labeled with T4 polynucleotide kinase using ã^{32}P-ATP.

Nucleases

Nucleases cleave phosphodiester bonds in the nucleic acids backbone. Based on their mode of action, two main classes have been defined: exonucleases are active at the end of nucleic acid molecules, and endonucleases cleave nucleic acids internally. Deoxyribonucleases cleave DNA and generates nicks (point in a double stranded DNA molecule where there is no phosphodiester bond between adjacent nucleotides of one strand, typically through damage or enzyme action), whereas ribonucleases cleave RNA. Nucleases are double edge swords for molecular biologists: on one hand, they are the worst enemy to nucleic acids integrity and, on the other hand, they are very useful to cut and manipulate nucleic acids for cloning purposes.

Deoxyribonucleases

Deoxyribonuclease I (DNase I, EC 3.1.21.1)

Deoxyribonuclease I (DNase I) is an endonuclease that acts on single- or double-stranded DNA (either isolated or incorporated in chromatin). DNAse I is used for nick translation of DNA, for generating random fragments for dideoxy sequencing, to digest DNA in RNA or protein preparations, and for DNA-protein interactions analysis in DNase footprinting.

DNase I is a 31 kDa glycoprotein, usually purified from bovine pancreas as a mixture of four isoenzymes (A, B, C, and D). It is important to note that crude preparations of DNAse I are often contaminated with RNase A. Thus, great attention should be paid to the quality control provided by the manufacturer to ensure lack of RNase A activity in DNase I preparations. At the end of the reaction, DNase I can be removed from the preparation by thermal denaturation at 75°C for 5 min in the presence of 5 mM EGTA (Huang et al. 1996).

DNase I-catalyzed cleavage occurs preferentially in 32 of a pyrimidine (C or T) nucleotide, and generates polynucleotides with free 32 -OH group and a 52 -phosphate. In the presence of Mg^{2+}, DNase I hydrolyzes each strand of duplex DNA independently, generating random cleavages. Maximal activity is obtained in the presence of Ca^{2+}, Mg^{2+}, and Mn^{2+} ions (Kunitz 1950). The nature of the divalent cations present in the incubation mixture affects both specificity and mode of action of DNase I. For instance, in the presence of Mn^{2+}, DNase I cleaves both DNA strands at approximately the same site, producing blunt ends or fragments with 1–2 base overhangs.

Exonuclease III (Exodeoxyribonuclease III, EC 3.1.11.2)

Exonuclease III was first isolated from the BE 257 *E. Coli* strain, which contains a thermo-inducible overproducing plasmid (pSGR3). Exonuclease III is a 28 kDa monomeric enzyme (Weiss 1976) with several interesting activities (Mol et al. 1995) 32 -exonuclease activity:

exonuclease III catalyzes the removal of mononucleotides from 32 - OH ends of double-stranted DNA; phosphatase activity: exonuclease III dephosphorylates DNA chains that terminate with a 32 -phosphate group (this type of chain is usually inert as a primer and inhibits DNA polymerase action); iii) RNase H activity: exonuclease III degrades RNA strands of DNA/RNA hybrids (Rogers and Weiss 1980) endonuclease activity: exonuclease III cleaves apurinic and apyrimidic bases from the sugar phosphate backbone in DNA (Keller and Crouch 1972). Another significant feature of exonuclease III is its relative specificity for double-stranted DNA. When this enzyme acts as an endonuclease, it generates a gap at nicks in the double-stranted DNA, whereas when it acts as an exonuclease, it generates a 52 -protruding end (that is resistant to further digestion since exonuclease III is not active on single-stranted DNA). Exonuclease III is unique among the exonucleases in its phospho-monoesterase action on a 32 -phosphate terminus. Exonuclease III is often used in conjunction with the Klenow fragment of *E. Coli* DNA polymerase to generate radio-labeled DNA strands. It is also used sequentially with S1 nuclease to reduce the length of double stranded DNA.

Optimum pH for exonuclease III's endonucleolytic and exonucleolytic activities is between 7.6 and 8.5, while phosphatase activity is maximal at pH between 6.8 and 7.4. The presence of Mg^{2+} or Mn^{2+} ions is required for optimal activity, while the presence of Zn^{2+} inhibits enzyme activity.

Bal 31 Nucleases (EC 3.1.11)

Two distinct molecular species described as fast (F) and slow (S) Bal 31 nucleases have been purified from the culture medium of *Alteromonas espejiana* Bal 31. Both species shorten duplex DNA (at both 32 and 52 ends) without introducing internal nicks, and exhibit a highly specific single stranded DNA endonuclease activity (cleaves at nicks, gaps and single-stranded regions of duplex DNA and RNA).

The purified F-Bal 31 nuclease is used for restriction mapping, removing long stretches (up to thousands of base pairs) or short stretches (tens to hundreds of base pairs) from duplex DNA, mapping B-Z DNA junctions, cleaving DNA at sites of covalent lesions (such as UV-induced), and shortening RNA molecules. The S-Bal 31 is a slower acting enzyme used only for restriction mapping and removal of short stretches from DNA duplex. Since Ca^{2+} is an essential cofactor, EGTA is required to inactivate both enzymes.

Most Bal 31-generated DNA fragments have fully base paired ends, which may further be ligated to any blunt end DNA fragments like molecular linkers. A small fraction of Bal 31-generated DNA fragments is harboring 52´-protruding ends, suggesting that Bal 31 acts sequentially as a 32´!52 exonuclease, followed by endonucleolytic removal of the protruding ends. Thus, the efficacy of a blunt-end ligation can be increased by utilizing Klenow fragment of *E. Coli* DNA polymerase or T4 DNA polymerase to fill the 52´-protuding ends generated by Bal 31.

Although F and S species of Bal 31 have similar activities on single stranded DNA, they act differently on double stranded DNA: F-Bal 31 can shorten duplex DNA approximately 20 times faster than S-Bal 31, although the reaction rate of digestion is dependent upon the C/G content of the substrate DNA. When tested under standard conditions, 1 ìg/ml of the F- and S-Bal 31 shortens DNA at rates of approximately 130 and 10 base pairs/terminus/minute, respectively.

Exonuclease VII (EC 3.1.11.6)

Exonuclease VII has first been isolated from the *E. Coli* K12 strain as a 88 kDa polypeptide. Exonuclease VII specifically degrades single stranded DNA, with no apparent activity on RNA or DNA/RNA hybrids. Since it is able to attack either end of the DNA molecules, it is often utilized to degrade long single protruding strands from duplex DNA (such as those generated by restriction endonucleases) and generate blunt end DNA. Exonuclease VII differs from exonucleases I and III in that i) it can degrade DNA from either its 52 or 32 end, ii) it can generate oligonucleotides, and iii) it is still fully active in 8 mM EDTA. Exonuclease VII is particularly useful for rapid removal of single-stranded oligonucleotide primers from a completed PCR reaction, when different primers are required for subsequent PCR reactions (Li et al. 1991).

Ribonucleases

Pancreatic Ribonuclease (RNase A, EC 3.1.27.5)

Pancreatic ribonuclease (also called RNase A) is an endonuclease that cleaves single stranded RNA at nucleoside 32´-phosphates and 32´-phospho-oligonucleotides ending in Cp or Up. RNase A is used to reduce RNA contamination in plasmid DNA preparations, and for mapping mutations in DNA or RNA by mismatch cleavage, since it cleaves the RNA in RNA/DNA hybrids at sites of single nucleotide mismatch.

RNAse A degrades the RNA into 32 -phosphorylated mononucleotides and oligonucleotides, generating a 22,32 -cyclic phosphate intermediate during the reaction.

RNase A is active under very different conditions and difficult to inactivate. Hence, great care needs to be taken to avoid contamination with RNAse A and further degradation of RNA samples. RNase-inhibitor from human placenta might be used to inhibit RNAse activity. However, we recommend using diethyl pyrocarbonate (DEPC), guanidinium salts, beta-mercaptoethanol, heavy metals, or vanadyl-ribonucleoside-complexes for efficient inhibition of RNAse A.

RNase A is available from many commercial sources. As stated above for other enzymes, very few suppliers provide useful information regarding origin or purity of their enzyme preparation.

Ribonuclease H (3.1.26.4)

Ribonuclease H (RNase H) specifically degrades RNA in RNA/DNA hybrids. It allows removal of RNA probes from prior hybridizations, or removal of poly-A tails at the 32 end of mRNAs.

Optimal activity of RNase H is achieved at a pH comprised between 7.5 and 9.1 (Berkower et al. 1973) in the presence of reducing reagents. RNase H activity is inhibited in the presence of N-ethylmaleimide (a chemical that react with SH groups), and is not markedly affected by high ionic strength (50% activity is retained in the presence of 0.3 M NaCl). RNase H requires Mg^{2+} ions, which can be replaced partially by Mn^{2+} ions.

Other Ribonucleases

Many other ribonucleases have been used for RNA sequencing. Each of them has specific cleavage requirement and specificity. Below is a list of these ribonucleases.

Ribonuclease Phy I can be used for rapid sequencing of RNA. It is isolated from cultures of *Physarum polycephalum*. Ribonuclease Phy I cleaves the RNA molecule at G, A, and U, but not at C residues. The products are 32 mononucleotides with 52 C termini.

RNase CL3 is used for sequencing RNA. It is isolated from chicken liver (*Gallus gallus*). RNase CL3 digests RNA adjacent to cytidilic acid in a ratio of 60 C residues digested for every U residue digested. The enzyme activity is inhibited by poly-A tracts, unless spermidine is added. Used primarily for RNA sequencing (Lockard et al. 1978),

Cereus ribonuclease is an endoribonuclease that preferentially cleaves RNA at U and C residues.

Ribonuclease Phy M, purified from Physarum polycephalum, has also been used for sequencing of RNA (Donis-Keller 1980). Ribonuclease Phy M preferentially cleaves RNA at U and A residues. Occasionally, cleavage at G may occur. This unwanted reaction can be minimized in the presence of 7 M urea.

RNase T1 is an endoribonuclease that specifically degrades single-stranded RNA at G residues. It cleaves the phosphodiester bond between 32 -guanylic residues and the 52 -OH residues of adjacent nucleotides with the formation of corresponding intermediate 22, 32 -cyclic phosphates. RNase T1 is an 11 kDa protein purified from *Aspergillus oryzae*.

RNAse T2 is also purified from *Aspergillus oryzae*. It has a molecular weight of 36,000, and cleaves all phosphodiester bonds in RNA, with a preference for adenylic bonds.

RNase U2, purified from *Ustilago sphaerogena*, is utilized as complement of RNase T1 in RNA sequencing, to discriminate purines residues since it specifically cleaves adenine residues when incubated at 50°C, pH3.5, and in the presence of 8 M urea. When incubated under standard conditions (Takahashi 1961), RNase U2 specifically cleaves the 32 -phosphodiester bond adjacent to purines, therefore generating purines 32 -phosphates or oligonucleotides with purine 32 -phosphate terminal groups. Cyclic 22,32 purine nucleotides are obtained as intermediates and that reversal of the final reaction step can be used to synthetize ApN and GpN. RNase U2 is thermostable (80°C for 4 min) in aqueous solution at pH 6.9.

DNA/RNA Nucleases

S1 Nuclease (EC 3.1.30.1)

S1 nuclease is a very useful tool for measuring the extent of hybridization (DNA— DNA or DNA— RNA), probing duplex DNA regions, and removing single stranded DNA of protruding ends generated by restriction enzymes. S1 nuclease is purified from *Aspergillus oryzae*. This enzyme degrades RNA or single stranded DNA into 52 mononucleotides, but does not degrade duplex DNA or RNA— DNA hybrids in native conformation. S1 nuclease is a 32 kDa metalloprotein (Vogt 1973). It requires Zn^{2+} for activity. Co^{2+} and Hg^{2+} can replace Zn^{2+}, but are less effective as cofactors. S1 nuclease's optimal activity is achieved at pH 4.0–4.3 (a 50% reduction of activity is observed at pH 4.9, and the enzyme is inactive at pH>6.0). S1 nuclease is strongly inhibited by chelating agents such as EDTA and citrate, or by low concentration of sodium phosphate (as low as 10 mM). Nuclease S1 is resistant to denaturing agents such as urea, SDS, or formamide and is thermostable (Ando 1966).

S1 nuclease hydrolyzes single-stranded DNA five times faster than RNA, and 75,000 times faster than double-stranded DNA. The low level of strand breaks which can be introduced by S1 nuclease in duplex DNA can be further reduced by high salt concentration (*i.e.* 0.2 M). S1 nuclease is active at S1 sensitive sites generated by negative supercoiling of the helical DNA structure, UV irradiation, or depurination (Shishido and Ando 1975).

S1 nuclease is widely used when specific removal of single portions of duplex DNA, RNA/DNA hybrids, or RNA molecules is required. Among these applications are mapping of spliced RNA molecules, isolation of duplex regions in single stranded viral genomes, probing strand breaks in duplex DNA molecules, cleavage of regions with lesser helix stability, localization of inverted repeated sequences, introduction of deletion mutation at D loop sites in duplex DNA, and mapping of the genomic regions involved in interactions with DNA binding proteins (Meyer et al. 1980).

Mung bean Nuclease

Mung bean nuclease can be used in the same way as nuclease S1, to remove protruding ends in duplex DNA or for transcription promoter mapping. Mung bean nuclease is a single stranded specific DNA and RNA endonuclease purified from mung bean sprouts. It yields 52 -phosphate terminated mono- and oligonucleotides. Complete duplex DNA degradation may occur when high enzyme concentration and extended incubation time are used. This is due to a two-step process: the enzyme first introduces single stranded nicks, followed by double stranded scissions and exonucleolytic digestion of the resulting fragments.

Proper trimming of the 52 -protruding extensions is achieved when the final blunt end contained a G-C base pair at its terminus (Ghangas and Wu 1975). Presence of an A-T base pair at the position where the fragment would end after trimming seemed to interfere with precise removal of the protruding end. Nucleotide composition of the overhang does not affect efficiency or quality of the nucleolytic digestion (Ghangas and Wu 1975). Mung bean nuclease was also proven to be very useful for excising cloned DNA fragments inserted in vectors following a dA·dT tailing (Wensink et al. 1974). Because poly(dA/dT) are not recognized as typical double stranded structures (Johnson and Laskowski 1970), they are hydrolyzed at half the rate of single stranded tails, but they are more efficiently cleaved than other duplex regions in DNA.

Mung bean nuclease requires Zn^{2+} and a reducing agent such as cysteine for maximum activity and stability. It is inhibited by high salt concentrations (80 to 90% inhibition in 200–400 mM NaCl). Triton X-100 (0.001% w/v) can increase mung bean nuclease stability (when used at low concentrations such as less than 50 U/il) and to prevent nuclease adhesion to surfaces.

Methylases

Methylases from Bacterial Restriction-modification Systems

Bacteria use restriction endonucleases to degrade foreign DNA introduced by infectious agents such as bacteriophages. In order to prevent destruction of its own DNA by the restriction enzymes, bacteria mark their own DNA by adding methyl groups to it (methylation). This modification, which must not interfere with the DNA base-pairing, usually affects only a few specific bases on each strand. Methylases are part of the bacterial restriction- modification (RM) system.

They catalyze the transfer of methyl groups from S-adenosyl-methionine (SAM) to specific nucleotides of double stranded DNA molecules. Methylases from type II RM systems (the most common) are encoded by separate proteins and act independently of their respective restriction endonucleases, whereas methylase and restriction activities of type I and III RM systems are provided by a unique protein complex (Wilson 1991).

Methylations often inhibit restriction enzymes that recognize the corresponding sequences (Sistla and Rao 2004), although there are exceptions to this rule (Gruenbaum et al. 1981a) some restriction endonucleases cleave DNA at a recognition sequence being modified by the *dam* or *dcm* methylases some bacteria have restriction endonucleases that only degrade methylated DNA and not the host unmethylated DNA (to overcome the camouflage evolution of some bacteriophages that contain methylated DNA).

Methylases are often used in molecular cloning to protect DNA from digestion, during DNA cloning or cDNA or genomic library construction. For example, when cloning a DNA fragment at the BamH I site of a given vector, the BamHI methylase can first be used to protect potential internal Bam HI sites in the vector, before addition of Bam HI linkers and digestion with the Bam HI endonuclease.

A methylase can also inhibit a restriction endonuclease of different RM systems (*e.g.* TaqI methylase inhibits BamHI restriction enzyme). The extent of overlap between restriction and methylation sequences determines the extent to which methylation alters endonuclease cleavage specificity and/or activity.

There are two types of overlaps: in the first case, the methylation sequence completely overlaps with the one of restriction endonuclease, and methylation alters cleavage specificity. In the second case, the methylation sequence partially overlaps with the one of restriction endonucleases, and nucleotide methylation modifies the endonuclease recognition sequence, which becomes resistant to the endonuclease activity.

The first class of overlap occurs when restriction endonucleases have degenerated recognition sequence, and when methylases are active on only one of the possible sites. For example, the Hinc II recognition sequence is GTPyPuAC, which means that Hinc II can cleave duplex DNA at the following four combinations: GTCGAC, GTCAAC, GTTGAC, and GTTAAC.

TaqI methylase (designated M.TaqI) catalyzes the transfer of a methyl group to the adenine residue of the TCGA sequence. Thus, among the four possible Hinc II recognition sites, only those containing a TCGA sequence are methylated by M.TaqI, and consequently become resistant to Hinc II digestion. This kind of overlap can be used to create new specificities for restriction endonucleases in duplex DNA (Nelson et al. 1984).

The second class of overlap occurs at the boundaries of the recognition sequences for a restriction endonuclease and a methylase. For example, when the GGATCC sequence of a Bam HI restriction site is followed by GG, the Bam HI site partially overlaps with the CCGG methylation site of M.MspI. Since M.MspI transfers a methyl group to the 52 cytosine residue in the CCGG sequence, it methylates the Bam HI site at its internal cytosine, thereby making this sequence resistant to the Bam HI endonuclease.

Dam and Dcm Methylases

Most B, K, and W strains of *E. Coli* also contain two site-specific DNA methylases, which are not part of the RM system and are encoded by the *dam* and *dcm* genes (Pirrotta 1976). The *dam* methylase transfers a methyl group from SAM to the N6 position of the adenine residue in the sequence GATC, while the *dcm* methylase (also known as *mec*) transfers a methyl group from SAM to the internal cytosine residues in the sequences CCAGG or CCTGG.

Dam methylation regulates post-replication mismatch repairs. All DNA isolated from *E. Coli* is not methylated to the same extent. For instance, the pBR322 plasmid DNA purified from *E. Coli* after amplification with chloramphenicol is resistant to the digestion by Mbo I, which does not cut at methylated sites. Furthermore, only 50% of the ë DNA obtained from a *dam*⁺ strain of *E. Coli* is methylated. The degree of *in vivo* cytosine methylation in the sequence CpG is related to the level of gene expression in eukaryotic cells, and quantity of methylated cytosine residues inversely correlates with gene activity.

Practical Considerations on the in Vitro use of Methylases

DNA fragments obtained by restriction endonuclease digestion of *in vitro* methylated DNA are in most respects indistinguishable from unmethylated DNA. However, it is important to note that methylated cytosine does not generate a band in the C channel when sequencing DNA by the Maxam and Gilbert method. In addition, the presence of methyl-cytosine residues in DNA reduces efficiency of transformation in most common strains of *E. Coli*.

This second problem results from the existence of restrictions systems in *E. Coli* K12 responsible for a specific degradation of DNA containing methylated cytosines. These restriction systems, designated 5-methylcytosine-specific restriction enzyme (*mcr*)-*A* and *mcrB*, are similar to those degrading hydroxymethylcytosine-containing DNA (*rglA* and *rglB*). The *mcrA*⁺ strains cleaves DNA modified by the Hpa II methylase while *mcrB* cleaves DNA modified by the Hae III, Alu I, Hha I, and Msp I methylases. *E. Coli* strains deficient in the *mcrB* system are available from New England Biolabs. DNA methylases from of type II RM systems perform the methylation reaction under similar conditions as do restriction endonucleases, except that the methylase requires SAM as a methyl group donor. Therefore, it is generally acceptable to carry out the methylation reaction using standard restriction endonuclease buffers to which SAM has been added. Importantly, methylases do not require divalent cations to be active, while most restriction endonucleases do.

Other Enzymes

Polyadenylate Polymerase of E. Coli (Polynucleotide Adenylyl-transferase, EC 2.7.7.19)

Polyadenylate (polyA) polymerase of *E. Coli* is a 58 kDa polypeptide that polymerizes adenylate residues at the 32 end of various polyribonucleotides, in a template-independent manner. *In vitro*, it is used for RNA 32 extension to prepare a priming site for cDNA synthesis using oligo-dT, or to prepare RNA for poly-dT-based purification. Depending on the experimental conditions, polyA polymerase can polymerize a polyA stretches from 20- to 2,000-nucleotide long.

E. Coli polyA polymerase catalyzes the addition of adenosine monophosphates (AMPs) using ATP as a substrate. ADP, dATP, and GTP are not polyA polymerase substrates, but CTP and UTP are (though at less than 5% of the rate obtained with ATP). PolyA polymerase is inhibited by phosphate and pyrophosphate ions, and aurin tricarboxylic acid, an inhibitor of both RNA polymerase and of the binding of mRNA to ribosomes. PolyA polymerase is a rather unusual enzyme in that its optimal activity requires the presence of high concentrations of monovalent cations (*e.g.* 400 mM NaCl). It is stimulated by Mn^{2+} and is insensitive to antibiotics such as rifampicin and streptolygdin (transcriptional inhibitors of initiation and elongation, respectively). In contrast to polynucleotide phosphorylase, another template-independent polynucleotide synthesizing enzyme, polyA polymerase does not degrade its own polyA product.

PolyA polymerase uses a wide variety of single stranded RNA species as primers. Double stranded RNA and some polynucleotides (such as polyUG, polyC, or di- and trinucleotides) are very poor primers. DNA does not function as a primer. This latter property is common to virtually all other polyA polymerases, except for the enzyme isolated from plants such as maize that shows considerable activity with natural and synthetic oligo- and poly-dNTPs.

Topoisomerase I (EC 5.99.1.2)

Topoisomerase I is a 105 kDa protein (Liu and Miller 1981) that catalyzes the relaxation of supercoiled DNA by transiently cleaving one strain of duplex DNA in the sugar— phosphate backbone, and further ligating the two generated ends. Topoisomerase I is an ubiquitous nuclear protein that is used by cells during processes such as replication, recombinaison, or DNA- protein interactions that involve unwinding of the DNA, and thereby create excessive supercoiling upstream of the unwinding point (Champoux 2001).

In vitro, topoisomerase I was initially used for circular DNA preparation (Martin et al. 1983), and for studying nucleosome assembly or DNA tertiary structures. Later, it was shown that topoisomerase I also catalyzes the covalent transfer of a single stranded DNA (donor) to an heterologous DNA (acceptor). This reaction involves the initial binding of the enzyme to single stranded DNA and formation of a covalent DNA/enzyme complex, which is then able to ligate the single stranded fragment to the 52 -OH end of the acceptor DNA. This ability of topoisomerase I to function as both a nuclease and a ligase has been utilized to develop new cloning vectors (Invitrogen TOPO® Cloning Vectors). The TOPO® cloning method uses topoisomerase I of *vaccinia* virus that i) specifically cuts double stranded DNA at the end of a (C/T)CCTT sequence, and ii) binds covalently to the 32 -phosphate of the thymidine of the generated DNA. The ligase activity of topoisomerase I can then join an acceptor DNA with compatible ends. TOPO® vectors form Invitrogen are provided as linear duplex DNA containing a topoisomerase I covalently linked to the 32 -phosphate on one strand, and a GTGG overhang on the other strand. Cloning is achieved using PCR products that have been amplified using a forward primer that contains four extra bases (CACC) at the 52 end. The overhang in the cloning vector (GTGG) hybridizes the 52 end of the PCR product, anneals to the added bases, and topoisomerase I ligates the PCR product in the correct orientation. Topoisomerase I also ligates the PCR product at the blunt 32 end.

Topoisomerase I is active at pH 7.5 in the presence of 50 to 200 mM NaCl, and is inactivated by 0.2% SDS. Covalent binding to DNA is stimulated by 5–10 mM Mg^{2+}.

Topoisomerase II (E. Coli DNA gyrase, EC 5.99.1.3)

Topoisomerase II, which is purified from *M. Luteus* (Klevan and Wang 1980), catalyzes the breakage and resealing of both strands of duplex DNA, thereby changing the linking number of discrete supercoiled forms of DNA by two. Topoisomerase II is also capable of reversibly knotting intact circular DNA. Type II topoisomerases are multimeric proteins that require ATP to be fully active. Type II topoisomerase is inhibited by etoposide, a chemotherapeutic agent used to reduce growth of rapidly dividing cancer cells (Baldwin and Osheroff 2005).

Guanylyl Transferase from Vaccinia Virus

Guanylyl transferase is a capping enzyme complex isolated from *Vaccinia* virus. Guanylyl transferase is used to label either 52 di- and triphosphate ends of RNA molecules, or capped 52 ends of RNA after chemical removal of the terminal 7-methyl-guanosine (m^7G) residue. This complex has three enzymatic activities: 1) it acts as an RNA triphosphatase by catalizing the cleavage of a pyrophosphate from a trisphosphate end of an RNA molecule, releasing a bisphosphate RNA end; 2) it is a guanylyl transferase, able to transfer one molecule of GTP to RNA, releasing pyrophosphate (because the 52 end of the RNA molecule ends in a phosphate group, the bond formed between the RNA and the GTP molecule is an unusual 52 -52 triphosphate linkage, instead of the 32 -52 bounds that exist between the other nucleotides forming an RNA strand); and 3) it is an RNA (guanine-7)-methyl transferase that transfers a methyl group from SAM to the 52 guanidine residue of an RNA molecule.

Guanylyl transferase complex does not accept monophosphate RNA as a substrate. Consequently, degraded or nicked RNA will not be labeled except at the 52 -cap end. The optimal amount of enzyme to be used must be determined empirically. During labelling experiments, it is recommended to verify that all labelling is incorporated in an authentic cap structure. To this end, tobacco acid pyrophosphatase can be used on an aliquot of the labeled samples: all labelling should be released in the form of 52 GMP after digestion.

7

Photosynthesis:
The Light Reactions

Principles of Spectrophotometry

Much of what we know about the photosynthetic apparatus was learned through spectroscopy—that is, measurements of the interaction of light and molecules. Spectrophotometry is an important branch of spectroscopy that focuses on the technique of measurement. Here we will examine four topics: Beer's law, the measurement of absorbance, action spectra, and difference spectra.

Beer's Law

An essential piece of information about any molecular species is how much of it is present. Quantitative measures of concentration are one of the cornerstones of biological science. Of all the methods that have been devised for measuring concentration, by far the most widely applied is absorption spectrophotometry. In this technique, the amount of light that a sample absorbs at a particular wavelength is measured and used to determine the concentration of the sample by comparison with appropriate standards or reference data. The most useful measure of light absorption is the absorbance (A), also commonly called the optical density (OD). The absorbance is defined as $A = \log I_0 / I$ where I_0 is the intensity of light that is incident on the sample and I is the intensity of light that is transmitted by the sample.

The absorbance of a sample can be related to the concentration of the absorbing species through Beer's law:

$$A = \varepsilon\, cl$$

where c is concentration, usually measured in moles per litter; l is the length of the light path, usually 1 cm; and ε is a proportionality constant known as the molar extinction coefficient, with the units of litres per mole per centimetre. The value of ε is a function of both the particular compound being measured and the wavelength. Chlorophylls typically have an ε value of about 100,000 L mol^{-1} cm^{-1}. When more than one component of a complex mixture absorbs at a given wavelength, the absorbances due to the individual components are generally additive.

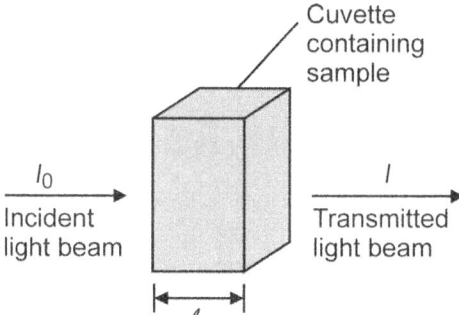

Figure 1: Definition of absorbance. A monochromatic incident light beam of intensity I_0 traverses a sample contained in a cuvette of length (l). Some of the light is absorbed by the chromophores in the sample, and the intensity of light that emerges is I.

The Spectrophotometer

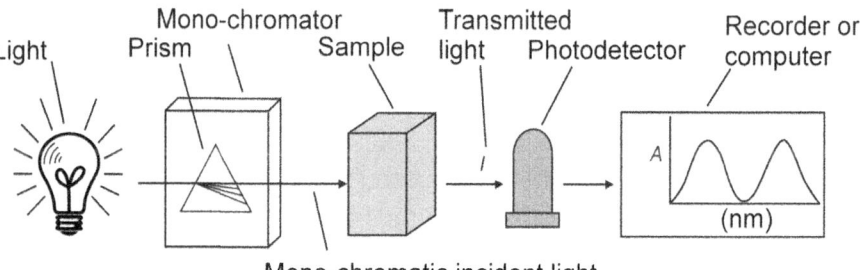

Figure 2: Schematic diagram of a spectrophotometer. The instrument consists of a light source, a monochromator that contains a wavelength selection device such as a prism, a sample holder, a photodetector, and a recorder or computer. The output wavelength of the monochromator can be changed by rotation of the prism; the graph of absorbance versus wavelength is called a spectrum.

The absorbance is measured by an instrument called a spectrophotometer. The essential parts of a spectrophotometer include a light source, a wavelength selection device such as a monochromator or filter, a sample chamber, a light detector, and a readout device, usually also include a computer, which is used for storage and analysis of the spectra. The most useful machines scan the wavelength of the light that is incident on the sample and produce, as output, spectra of absorbance versus wavelength.

Action Spectra

The use of action spectra has been central to the development of our current understanding of photosynthesis. An action spectrum is a graph of the magnitude of the biological effect observed as a function of wavelength. Examples of effects measured by action spectra are oxygen evolution and hormonal growth responses due to the action of phytochrome. Often an action spectrum can identify the chromophore responsible for a particular light-induced phenomenon. Action spectra were instrumental in the discovery of the existence of the two photosystems in O_2-evolving photosynthetic organisms.

Some of the first action spectra were measured by T. W. Engelmann in the late 1800s. Engelmann used a prism to disperse sunlight into a rainbow that was allowed to fall on an aquatic algal filament. A population of O_2-seeking bacteria was introduced into the system. The bacteria congregated in the regions of the filaments that evolved the most O_2. These were the regions illuminated by blue light and red light, which are strongly absorbed by chlorophyll. Today, action spectra can be measured in room-sized spectrographs in which the scientist enters a huge monochromator and places samples for irradiation in a large area of the room bathed by monochromatic light. But the principle of the experiment is the same as that of Engelmann's experiments.

Schematic diagram of the action spectrum measurements by T. W. Engelmann. Engelmann projected a spectrum of light onto the spiral chloroplast of the filamentous green alga *Spirogyra* and observed that oxygen-seeking bacteria introduced into the system collected in the region of the spectrum where chlorophyll pigments absorb. This action spectrum gave the first indication of the effectiveness of light absorbed by accessory pigments in driving photosynthesis.

Difference Spectra

An important technique in studies of photosynthesis is light-induced difference spectroscopy, which measures changes in absorbance. In this technique, bright light, often called actinic light, is used to illuminate a sample, while a dim beam of light is used to measure the absorbance of the sample at wavelengths other than that of the actinic beam. In this way a difference spectrum is obtained, which represents the changes in the absorption spectrum of the sample induced by illumination with the actinic light. Absorption bands that disappear upon illumination appear as negative peaks; new bands that appear upon illumination appear as positive peaks. Difference spectra give important clues to the identity of molecular species participating in the photoreactions of photosynthesis.

By the use of special flash techniques, it is possible to record the difference spectrum at a given time after flash excitation. Multiple difference spectra recorded at different times after flash excitation can be used to measure the kinetics of the chemical reactions that follow photon excitation of a reaction centre. These techniques can have extraordinary time resolution, in some cases less than a picosecond (10^{-12} s), and have provided great insights into the earliest events in the photosynthetic energy storage process.

Antagonistic Effects of Light on Cytochrome Oxidation

The puzzling red drop and enhancement effects were explained by experiments performed by Louis Duysens of the Netherlands. He suggested that two phtotchemical events were responsible for enhancement effects. One photochemical event produced an oxidation while the other produced a reduction. Chloroplasts contain cytochromes, iron-containing proteins that function as intermediate electron carriers in photosynthesis. Duysens found that when a sample of a red alga was illuminated with long-wavelength light, the cytochrome became mostly oxidized. If light of a shorter wavelength was also present, the effect was partly reversed. These antagonistic effects can be explained by a mechanism involving two photochemical events: one that tended to oxidize the cytochrome and one that tended to reduce it.

We know now that in the red region of the spectrum, one of the photoreactions, known as photosystem I (PS-I), absorbs preferentially far-red light of wavelengths greater than 680 nm, while the second, known as photosystem II (PS-II), absorbs red light of 680 nm well and is driven very poorly by far-red light. This wavelength dependence explains the enhancement effect and the red drop effect. Another difference between the photosystems is that Photosystem I produces a strong reductant, capable of reducing NADP⁺, and a weak oxidant. Photosystem II produces a very strong oxidant, capable of oxidizing water, and a weaker reductant than the one produced by photosystem

I. This reductant of photosystem II rereduces the oxidant produced by photosystem I, which explains the antagonistic effect.

Structures of Two Bacterial Reaction Centres

In 1984, Hartmut Michel, Johann Deisenhofer, Robert Huber, and coworkers in Munich solved the three-dimensional structure of the reaction centre from the purple photosynthetic bacterium *Rhodopseudomonas viridis*. This landmark achievement, for which a Nobel prize was awarded in 1988, was the first high-resolution X-ray structure determination for an integral membrane protein and the first structure determination for a reaction centre complex.

The protein part of the complex consists of four separate polypeptides. Two of them, called L and M (for *l*ight and *m*edium mass) bind all of the bacteriochlorophyll, quinone, and carotenoid cofactors of the complex. The structure has a twofold symmetry about an axis perpendicular to the plane of the membrane, hinting at a dimeric nature of the reaction centre. The ten transmembrane portions of the L and M peptides (five from each) are arranged in α helices, and there are almost no charged amino acid residues in the interior of the membrane. The H (*h*eavy) protein has a single transmembrane helix and is localized mostly on the cytoplasmic side of the membrane. The C (*c*ytochrome) subunit is located in the periplasmic region (the region between the bacterial plasma membrane and the outer membrane). The geometric arrangement of the pigments and the quinones (electron acceptors), with the protein removed. A similar arrangement is found in the reaction centre of another purple photosynthetic bacterium, *Rhodobacter sphaeroides*, except that the C subunit is not present.

Two of the bacteriochlorophyll molecules are in intimate contact with each other and are known as the special pair. This dimer, whose existence was predicted from magnetic-resonance studies, is the photoactive portion of the complex. An electron is transferred from this dimer along the sequence of electron carriers on the right side of the complex. Detailed analysis of these structures, along with analysis of numerous mutants, has revealed many of the principles involved in the energy storage processes that are carried out by all reaction centres.

The bacterial reaction centre structure is thought to be similar in many ways to that found in photosystem II from oxygen-evolving organisms, especially in the electron acceptor portion of the chain. The proteins that make up the core of the bacterial reaction centre are relatively similar in sequence to their photosystem II counterparts, implying an evolutionary relatedness.

Midpoint Potentials and Redox Reactions

Redox reactions, midpoint potentials, and their relationship to the laws of thermodynamics. These concepts are useful for our discussion of electron flow from H_2O to $NADP^+$ and the interactions between the different electron carriers.

The midpoint potential (E_m) is a measure of the tendency of a compound to take electrons from other compounds. A large positive midpoint potential means that the compound is a strong oxidant; a large negative value means that the compound is a strong reductant (in both cases relative to the standard hydrogen electrode).

Equilibrium constants can easily be predicted from midpoint potentials, in the same way that free energies were related to equilibrium constants. Midpoint potentials for many chemical and biochemical reactions have been measured and tabulated. The y-axis on the Z scheme, midpoint potentials of the electron carriers, with negative values higher than positive ones. This choice makes reactions that are spontaneous (releasing free energy) appear "downhill" on the graph.

Knowledge of the midpoint potentials of the various electron carriers is important in establishing the pathway of electron flow in any biochemical electron transport system, such as those found in chloroplasts or mitochondria. Researchers make this measurement usually by carrying out a redox titration. They adjust, or poise, the sample at a particular redox potential, usually by adding small amounts of oxidants or reductants. Redox mediators, small molecules that permit rapid equilibration between the sample and the electrodes of the measurement system, must be included to ensure that the system is at equilibrium when the measurement is made. Several measurements are made at a variety of redox potentials. The sample is stirred in a special cell that contains platinum and reference electrodes, and chemical oxidants and reductants are added to adjust the redox potential, which is read with a voltmeter.

We can measure the extent of the redox reaction by following a particular property of the sample, usually absorbance, at each potential. In the example illustrated, the reduced form of the compound has an absorbance that decreases as the compound is oxidized. The fraction of reduced form at each potential is plotted against redox potential, and the midpoint potential (E_m) is determined as the potential at which the compound is half oxidized and half reduced.

Oxygen Evolution

The chemical mechanism of photosynthetic water oxidation is not yet known, although there is a great deal of indirect evidence about the process. If a sample of dark-adapted photosynthetic membrane is exposed to a sequence of very brief, intense flashes, a characteristic pattern of oxygen production is observed. Little or no oxygen is produced on the first two flashes, and maximal oxygen is released on the third flash and every fourth flash thereafter, until eventually the yield per flash damps to a constant value. This remarkable result was first observed by Pierre Joliot in the 1960s.

A schematic model explaining these observations, proposed by Kok and coworkers, has been widely accepted. This model for the photooxidation of water, called the S state mechanism, consists of a series of five states, known as S_0 to S_4, which represent successively more oxidized forms of the water-oxidizing enzyme system, or oxygen-evolving complex. The light flashes advance the system from one S state to the next, until state S_4 is reached. State S_4 produces O_2 without further light input and returns the system to S_0. Occasionally, a centre does not advance to the next S state upon flash excitation, and less frequently, a centre is activated twice by a single flash. These misses and double hits cause the synchrony achieved by dark adaptation to be lost and the oxygen yield eventually to damp to a constant value. After this steady state has been reached, a complex has the same probability of being in any of the states S_0 to S_3 (S_4 is unstable and occurs only transiently), and the yield of O_2 becomes constant. States S_2 and S_3 decay in the dark, but only as far back as S_1, which is stable in the dark. Therefore, after adaptation to the dark, approximately three-fourths of the oxygen-evolving complexes appear to be in state S_1 and one-fourth in state S_0. This distribution of states explains why the maximum yield of O_2 is observed after the third of a series of flashes given to dark-adapted chloroplasts.

Photosystem I

The PS-I reaction centre is composed of a multiprotein complex. The reaction centre chlorophyll P700 and about 100 core antenna chlorophylls are bound to two proteins, PsaA and PsaB, with molecular masses in the range of 66 to 70 kDa.

PS-I reaction centre complexes have been isolated from several organisms and found to contain the 66 to 70 kDa proteins, along with a variable number of smaller proteins in the range of 4 to 25 kDa. Some of these proteins serve as binding sites for the soluble electron carriers plastocyanin and ferredoxin. The functions of some of the other proteins are not well understood. An 8-kDa protein contains some of the bound iron-sulfur centres that serve as early electron acceptors in photosystem I. The structure of the PS-I complex from pea has been determined to a resolution of 4.4 E, and the positions of many of the chlorophylls and electron transfer components have been located.

In their reduced form, the electron carriers that function in the acceptor region of photosystem I are all extremely strong reducing agents. These reduced species are very unstable and thus difficult to identify. Evidence indicates that one of these early acceptors is a chlorophyll molecule, and another is a quinone species, phylloquinone, also known as vitamin K_1.

Additional electron acceptors include a series of three membrane-associated iron–sulfur proteins, or bound ferredoxins, also known as Fe–S centres Fe–S_X, Fe–S_A, and Fe–S_B. Fe–S centre Fe–S_X is part of the P700-binding protein; centres Fe–S_A and Fe–S_B reside on an 8 kDa protein that is part of the PS-I reaction centre complex. Electrons are transferred through centres Fe–S_A and Fe–S_B to ferredoxin, a small, water-soluble iron–sulfur protein. The membrane-associated flavoprotein ferredoxin–NADP reductase (FNR) reduces $NADP^+$ to NADPH, thus completing the sequence of noncyclic electron transport that begins with the oxidation of water. Electron transport between PSII and PSI is mediated by plastohydroquinone, the cytochrome b_6f complex, and plastocyanin. The cytochrome b_6f complex contains two b-type hemes and one c-type heme.

A model for the organization of electron carriers in PS-I is shown in Figure below. The P700 dimer is located at the bottom of the structure and two symmetrical arms radiate from P700. Each arm includes an accessory chlorophyll a and another chlorophyll molecule tentatively identified as A_0. Other structures include the Fe-S centre F_X, and two other Fe-S centres, F_1 and F_2. The distance between the carriers is shown at the right.

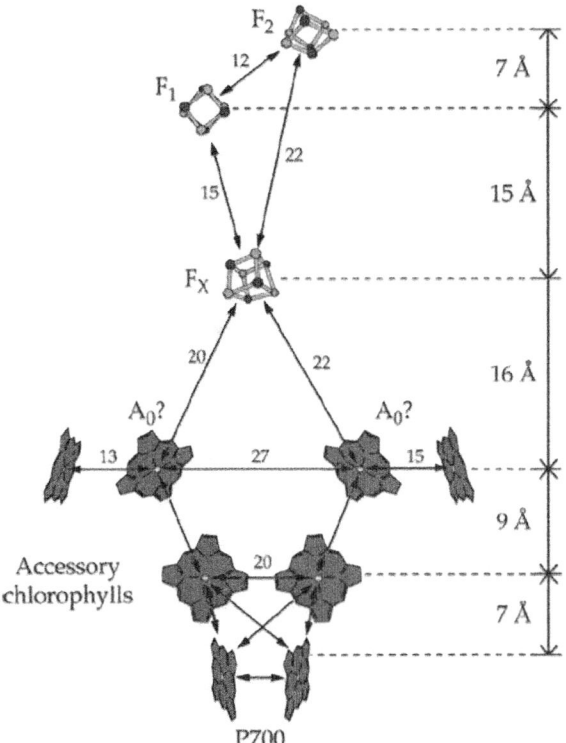

Figure: A model for the organization of electron carriers in PS-I.

The PS-I reaction centre appears to have some functional similarity to the reaction centre found in the anaerobic green sulfur bacteria and the heliobacteria. These bacteria contain low-potential Fe–S centres as early electron acceptors and are probably capable of ferredoxin-mediated NAD^+ reduction similar to the $NADP^+$ reduction function of photosystem I. There is almost certainly an evolutionary relationship between these complexes and photosystem I of oxygen-evolving organisms.

ATP Synthase

The chemical mechanism by which ATP is synthesized is not yet understood in detail. However, considerable evidence now supports a mechanism, first proposed by Paul Boyer, in which the principal energy-requiring step is the release of bound ATP from the enzyme.

It has also been proposed that during catalysis a large portion of the CF_1 complex rotates about a bearing consisting of the γ subunit.

The γ subunit may act as a camshaft does, rotating alternately against the α and α subunits. The energy of the conformational movements is then translated into phosphoanhydride bond energy.

The thylakoid ATP synthase has two segments: CF_0 is a transmembrane segment, and CF_1 is a hydrophilic segment at the stomal surface. CF_0 traslocates protons from the chloroplast lumen to the catalytic part of the enzyme and CF_1 cataylses the conversion of ADP and inorganic phosphate to ATP.

The chloroplast ATP synthase has nine different subunits. CF_1 is made of two large subunits, α and β, plus three smaller subunits, γ, δ, and ε. Each CF_1 molecule has three copies of the α and β subunits and one copy of the γ, δ and ε subunits. The α and β subunits bind ADP and phosphate and catalyize the phosphorylation of ADP into ATP.

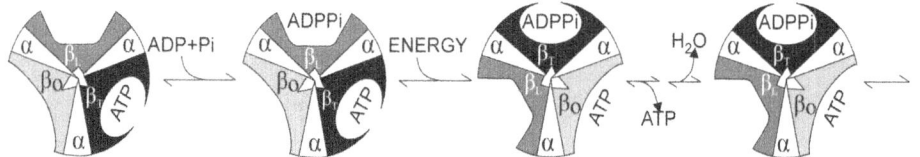

Figure: The model of Boyer's binding change mechanism. The a and b subunits configure three nucleotide binding sites: O, which provides the early binding site for ADP and inorganic phosphate, L, to which the ADP and inorganic phosphate bind after migrating from O, and T, which tightly binds ATP. Energy ensuing from the movement of protons from the chloroplast lumen to the stroma drives the rotation of the g subunit of CF_1, and the interconversion of the binding sites and the release of an ATP molecule.

Studies on the mitochondrial ATP synthase have increased our understanding of how the ATP synthases work. As proposed by Paul Boyer in 1997, ATP synthesis takes place by a binding change mechanism.

According to this model, the energy of the proton gradient is used to release a tightly bound form of ATP from a catalytic binding side of the enzyme. As shown in Figure below, CF_1 has three nucleotide binding sites, L, T and O. ADP and inorganic phosphate are postulated to first bind to the O site, and then move to the L site.

The T site binds ATP. As protons move from the lumen to the stromal region through the CF_0 channel, energy is released and the γ subunit of CF_1 rotates. The rotation causes conformation changes in the three nucleotide binding sites, which interconvert, thus changing the afiity of the sites for the nucleotides. As the T site converts into a O site, ATP is released and another cycle starts. Direct observation of the rotation of the γ subunit of CF$_1$ has provided strong experimental support for the binding change mechanism.

Mode of Action of Some Herbicides

Herbicides of one major class—about half of the commercially important compounds—act by interrupting photosynthetic electron flow.

A shows the chemical structure of two of these compounds. The precise sites of action of many of these agents have been found to lie either at the reducing side of photosystem I (for example, in paraquat) or in the quinone acceptor complex in the electron transport chain between the two photosystems.

Figure: The use of herbicides to kill unwanted plants is widespread in modern agriculture. Many different classes of herbicides have been developed, and they act by blocking amino acid, carotenoid, or lipid biosynthesis or by disrupting cell division. Understanding the mode of action of herbicides has been an important tool in research on plant metabolism and has facilitated their application under different agricultural practices (Ashton and Crafts 1981). Here, the chemical structure of two herbicides that block photosynthetic electron flow are shown. DCMU is also known as diuron. Paraquat has acquired public notoriety because of its use on marijuana crops.

Paraquat acts by intercepting electrons between the bound ferredoxin acceptors and NADP and then reducing oxygen to superoxide (O_2^-). Superoxide is a free radical that reacts nonspecifically with a wide range of molecules in the chloroplast, leading to the rapid loss of chloroplast activity. Lipid molecules in cell membranes are especially sensitive.

The herbicides that act on the quinone acceptor complex compete with plastoquinone for the Q_B binding site. If herbicide is present, it displaces the oxidized form of plastoquinone and occupies the specific binding site for the quinone acceptor, which is thought to lie on the D1 herbicide-binding protein. The herbicide is not able to accept electrons, so the electron is unable to leave Q_A, the first quinone acceptor. Thus, the binding of herbicide effectively blocks electron flow and inhibits photosynthesis. Many herbicides that act in this manner also inhibit electron flow in photosynthetic bacteria that have quinone-type electron acceptor complexes.

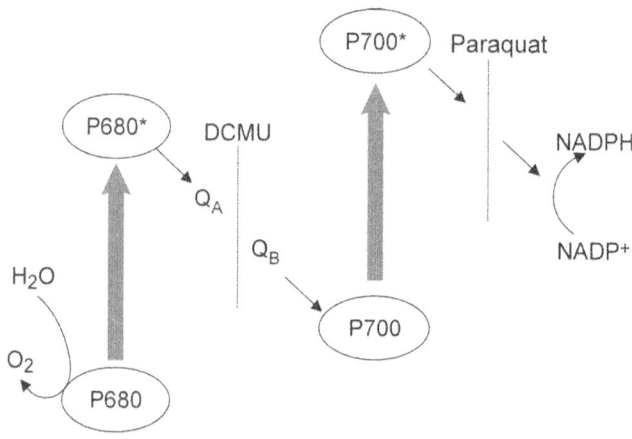

Figure: Sites of action of herbicides. Many herbicides, such as DCMU, act by blocking electron flow at the quinone acceptors of photosystem II, by competing for the binding site of plastoquinone that is normally occupied by Q_B. Other herbicides, such as paraquat, act by accepting electrons from the early acceptors of photosystem I and then reacting with oxygen to form superoxide, O_2^-, a species that is very damaging to chloroplast components, especially lipids.

In recent years, herbicide-resistant biotypes of common weeds have appeared in areas where a single type of herbicide has been used continuously for several years. These biotypes can be orders of magnitude more resistant to certain classes of herbicides than the nonresistant plant is. In several cases the resistance factor has been traced to a single amino acid substitution in the D1 protein. This change, presumably in the quinone (and herbicide) binding region of the peptide, lowers the binding affinity of the herbicide, making it much less effective.

The possibility of fine-tuning the herbicide sensitivity of crop plants by making subtle changes in the proteins of the PS-II reaction centre has created a great deal of interest in the agricultural chemical industry. Through biotechnology, it is now possible to make a crop plant that is resistant to a particular herbicide, which can then be applied to control undesirable plants that are not resistant. The success of this approach will depend on whether undesirable side effects of the herbicide-resistant mutations can be controlled and on how rapidly the weeds acquire resistance through natural selection or gene transfer. Thus, it is a race between science and evolutionary change.

A complementary approach is to insert resistance factors to certain pests, such as insects, into the plant via genetic engineering. The plants then produce the toxic species themselves, eliminating the need for insecticides.

Chlorophyll Biosynthesis

In the first phase of chlorophyll biosynthesis, the amino acid glutamic acid is converted to 5-aminolevulinic acid (ALA). This reaction is unusual in that it involves a covalent intermediate in which the glutamic acid is attached to a transfer RNA molecule. This is one of a very small number of examples in biochemistry in which a tRNA is utilized in a process other than protein synthesis. Two molecules of ALA are then condensed to form porphobilinogen (PBG), which ultimately form the pyrrole rings in chlorophyll. The next phase is the assembly of a porphyrin structure from four molecules of PBG. This phase consists of six distinct enzymatic steps, ending with the product protoporphyrin IX.

All the biosynthesis steps up to this point are the same for the synthesis of both chlorophyll and heme. But here the pathway branches, and the fate of the molecule depends on which metal is inserted into the centre of the porphyrin. If magnesium is inserted by an enzyme called magnesium chelatase, then the additional steps needed to convert the molecule into chlorophyll take place; if iron is inserted, the species ultimately becomes heme.

The next phase of the chlorophyll biosynthetic pathway is the formation of the fifth ring (ring E) by cyclization of one of the propionic acid side chains to form protochlorophyllide. The pathway involves the reduction of one of the double bonds in ring D, using NADPH. This process is driven by light in angiosperms and is carried out by an enzyme called protochlorophyllide oxidoreductase (POR). Non-oxygen-evolving photosynthetic bacteria carry out this reaction without light, using a completely different set of enzymes. Cyanobacteria, algae, lower plants, and gymnosperms contain both the light-dependent POR pathway and the light-independent pathway. Seedlings of angiosperms grown in complete darkness lack chlorophyll, because the POR enzyme requires light. These *etiolated* plants very rapidly turn green when exposed to light. The final step in the chlorophyll biosynthetic pathway is the attachment of the phytol tail, which is catalyzed by an enzyme called chlorophyll synthetase.

The elucidation of the biosynthetic pathways of chlorophylls and related pigments is a difficult task, in part because many of the enzymes are present in low abundance. Recently, genetic analysis has been used to clarify many aspects of these processes.

Photosynthesis: Carbon Reactions

How the Calvin Cycle Was Elucidated

The elucidation of the Calvin cycle was the result of a series of elegant experiments by Melvin Calvin and his colleagues in the 1950s, for which a Nobel prize was awarded in 1961. They used suspensions of the unicellular eukaryotic green alga *Chlorella* to trace the path of carbon. The Calvin cycle is also referred to as the reductive pentose phosphate cycle to distinguish the photosynthetic cycle from the oxidative pentose phosphate pathway, with which it shares several enzymes. Yet another designation of the Calvin cycle is photosynthetic carbon reduction cycle.

To elucidate the cycle, the investigators first exposed the algal cells to constant conditions of light and CO_2 to establish steady-state photosynthesis. Next they added radioactive $^{14}CO_2$ for a brief period to label the intermediates of the cycle. They then killed the cells and inactivated their enzymes by plunging the suspension into boiling alcohol. They separated the ^{14}C-labeled compounds from one another and identified them by their positions on two-dimensional paper chromatograms.

In this manner, 3-phosphoglycerate and the various sugar phosphates were identified as intermediates in carbon fixation. By exposing the cells to $^{14}CO_2$ for progressively shorter periods of time, Calvin's group was able to identify 3-phosphoglycerate as the first stable intermediate.

Therefore, the other labelled sugar phosphates must be derived from a subsequent reduction of 3-phosphoglycerate.

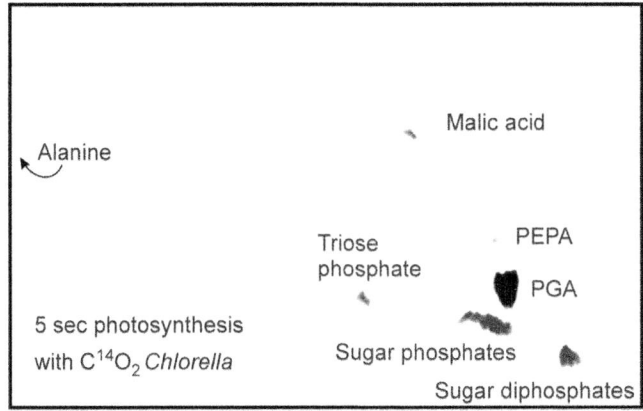

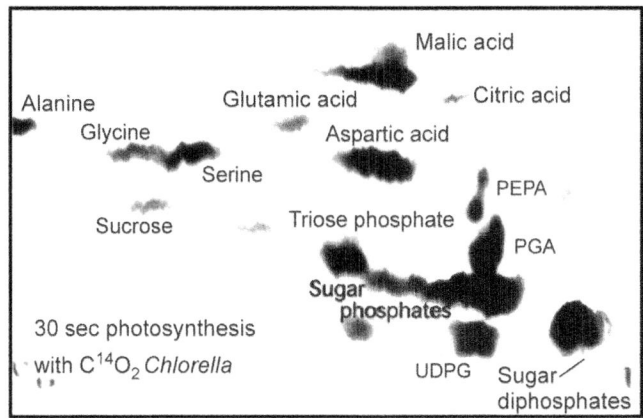

Figure: Autoradiograms showing the labelling of carbon compounds in the alga Chlorella after exposure to $^{14}CO_2$. The time intervals shown in the figures indicate the length of exposure to the radiolabel. At the indicated time intervals, the reaction was terminated, by plunging of the contents into boiling alcohol. The labelled compounds in the cell homogenates were then separated by paper chromatography. The heavy labelling of 3-phosphoglycerate (PGA) after the shorter exposure indicates that it is the first stable intermediate of the Calvin (reductive pentose phosphate) cycle.

To deduce the path of carbon, it was necessary to determine the distribution of ^{14}C in each of the labelled sugars. After a brief exposure to $^{14}CO_2$, individual intermediates were isolated and chemically degraded so that the amount of ^{14}C in each carbon atom could be determined. The results showed that 3-phosphoglycerate was initially labelled predominantly in the carboxyl group. This finding suggested that the initial CO_2 acceptor was a two-carbon compound, and it prompted a long and futile search for such a compound.

The subsequent discovery that pentose monophosphates and a pentose bisphosphate (ribulose) participate in the cycle raised the possibility that the initial CO_2 acceptor was a five-carbon compound. This conceptual breakthrough rapidly led to the identification of ribulose-1,5-bisphosphate as the CO_2 acceptor and to the formulation of the complete cycle. The operation of the cycle explained the ^{14}C-labeling pattern of the other intermediates. To demonstrate conclusively that a metabolic pathway exists, it is necessary to prove that the postulated enzymes catalyze the proposed reactions in the test tube. Ideally, the rates of these enzyme reactions (in vitro) should be equal to or in excess of those observed in the intact cell (in vivo). However, this evidence can be used only to support a proposed pathway. Failure to demonstrate a reaction in vitro does not prove that the reaction does not occur.

All of the reactions of the Calvin cycle have been demonstrated in vitro. With one exception, the in vitro rates of enzymatic activity are well in excess of the maximal observed rates of photosynthesis. That exception is rubisco, which, when assayed under the levels of CO_2 and O_2 in air, shows barely enough in vitro activity to account for the observed rate of photosynthesis in air. Although great progress has been made in increasing activity by providing proper conditions for activation, researchers continue to debate whether rubisco is a rate-limiting step in photosynthesis.

Rubisco: A Model Enzyme for Studying Structure and Function

With the exception of the carbon fixed by a some prokaryotic organisms, most of the carbon fixed on Earth is processed by the Calvin cycle. The concentration of rubisco, the carboxylating enzyme of the Calvin cycle, is generally high. For example, it accounts for 50% or more of the total protein in plant leaves, and its concentration within chloroplasts is extremely high (*ca.* 0.2 g/ml).

Rubisco occurs in two functionally analogous forms, having different structure, distribution, and O_2 sensitivity. Oxygenic phototroph (cyanobacteria, chloroplasts from high plants) and many photosynthetic bacteria contain a form of the enzyme, (form I), made up of eight large (L) catalytic subunits (about 55 kDa each) and eight small (S) subunits (about 14 kDa each), giving a molecular mass of about 560 kDa for the complete protein (L_8S_8). The 8 L subunits of form I are arranged as an octameric core surrounded by two layers of four S subunits, with each layer located on opposite sides of the molecule.

Form II of rubisco is found in some photosynthetic purple nonsulfur bacteria, and it is composed of two L subunits, each 50 kDa. Eukaryotic algal dinoflagellates also use a dimeric enzyme of proteobacterial ancestry. A rubisco-like enzyme lacking catalytic activity has recently been identified in photosynthetic green sulfur bacteria which lack the Calvin cycle. The evidence suggests that the enzyme may function in sulfur oxidation to provide electrons for the photosynthetic electron transport chain. Genome sequences show that the rubisco-like protein also occurs in certain heterotrophic bacteria.

Rubisco Activase

Rubisco catalyzes the incorporation of CO_2 to the C-2 of the enediol form of ribulose-1,5-bisphosphate and the subsequent cleavage of the unstable intermediate to yield two molecules of 3-phosphoglycerate The conversion of the inactive form of rubisco to an active state, which is required for both the carboxylation and oxygenation reactions, requires the carbamylation of a specific lysine residue (in spinach, Lys-201). As a consequence of the binding of the CO_2^- molecule to the lysine residue, the enzyme acquires a triad of anionic residues that provide the binding site for the cofactor Mg^{2+} (in spinach, the other two residues are Asp-203 and Glu-204). At this stage, the incorporation of ribulose-1,5-bisphosphate causes conformational changes in rubisco that allows the enzyme to sequester the sugar bisphosphate from the solvent and, upon the binding of a CO_2 or O_2 molecule, to start the catalytic cycle. However, the tight binding of ribulose-1,5-bisphosphate to the uncarbamylated enzyme molecules displaces the equilibrium to the dead end complex, rubisco-ribulose-1,5-bisphosphate, so that the rate of reaction declines rapidly until it finally stops. How is this problem solved?

A combination of genetic, physiological and biochemical studies have provided the solution to the long-term riddle of rubisco activation. In vivo, some mutants of *Arabidopsis thaliana* lack light-mediated activation of rubisco and can only grow at high concentrations of CO_2. Surprisingly, the mutation does not impair the response of the enzyme to modulation by CO_2 and Mg^{2+} in vitro. These contradictory results led to the discovery of a protein, rubisco-activase, that uncouples ribulose-1,5-bisphosphate from decarbamylated active sites and, in so doing, promotes the access of CO_2 and Mg^{2+} for the carbamylation of the enzyme. The importance of this process is that the rate at which rubisco-activase enhances the catalytic capacity of rubisco is linked to the hydrolysis of ATP (50 molecules of ATP hydrolyzed / 1 rubisco site activated).

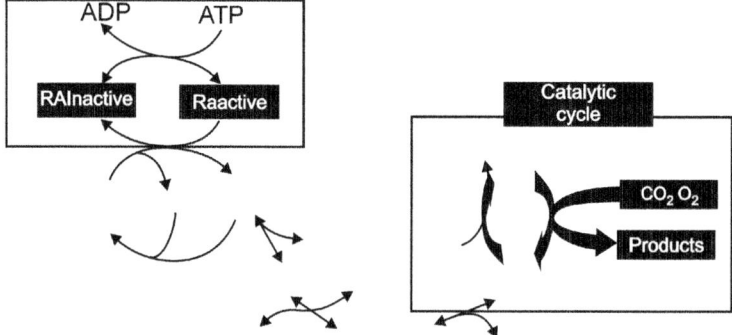

Figure: Modulation of rubisco by rubisco-activase. The combination or the interaction of rubisco (E) with modulators distributes the enzyme among different states (yellow boxes). The unprotonated form of a specific lysine residue (E-NH$_2$) reacts with CO$_2$ releasing a proton and forming the carbamate (E-NH-CO$_2$) that is stabilized by the cofactor Mg^{2+} (E-NH-CO$_2^-$. Mg^{2+}). At this stage (blue box), rubisco is catalytically competent and thereby binds ribulose-1,5-bisphosphate (RuBP) yielding the ternary complex (E-NH-CO$_2^-$. Mg^{2+}. RuBP) in which the latter is converted to the enediol form. This entails the incorporation of CO$_2$ or O$_2$ in the active site to generate the products of carboxylation or oxygenation, respectively. The high affinity of ribulose-1,5-bisphosphate for free rubisco (E-NH^{3+}) generates the dead-end complex (E-NH^{3+}. RuBP) thereby displacing the above equilibria to a catalytically incompetent state. Rubisco-activase (RA) interacts noncovalently with the binary complex and facilitates the release of ribulose-1,5-bisphosphate (red box) freeing the enzyme for the next carbamylation cycle. This process requires the concurrent hydrolysis of ATP.

Two isoforms of rubisco activase, (42 and 46 kDa) found in several species differ in the C-terminal region as result of the differential splicing of a common pre-mRNA. Both isoforms catalyze the activation of rubisco and the hydrolysis of ATP but the smaller isoform has higher capacity than the larger isoform for both activities.

Evidence for the fact that the two processes are independent are provided by studies showing that the digestion of native rubisco-activase with trypsin or the modification of its primary structure by site-directed mutagenesis do not elicit the same responses in the stimulation of rubisco and the hydrolysis of ATP. More importantly, the deletion of some residues at the C-region of rubisco-activase doubles the rate of rubisco activation with only a minor effect on the hydrolysis of ATP.

Rubisco activation by the activase requires a mutual recognition of both proteins. Rubisco-activase from tobacco (*Solanaceae*) fails to activate rubisco from spinach or *Chlamydomonas* (non-*Solanaceae*) and, similarly, the rubisco-activase from spinach or *Chlamydomonas* fails to stimulate rubisco from tobacco. However, the substitution of some residues on the surface of the large subunit of *Chlamydomonas* rubisco shifts the specificity from non-*Solanaceae* to *Solanaceae* rubisco-activase. These experiments underscore the contribution of specific amino acids to protein-protein interaction and provide experimental evidence for a recognition step separated from the subsequent activation process.

Three Variations of C$_4$ Metabolism

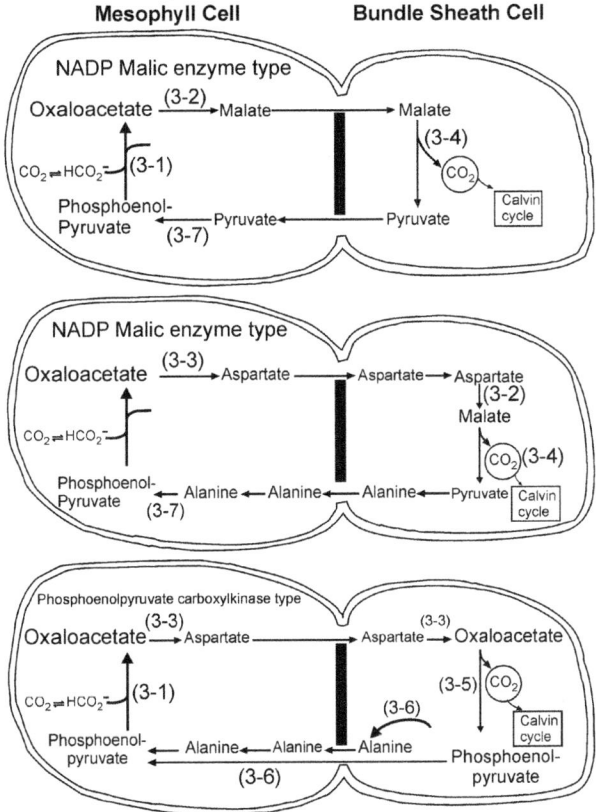

Figure: The three variants of the C$_4$ photosynthetic carbon cycle. The variants differ principally in (1) the nature of the four-carbon acid (malate or aspartate) transported into the bundle sheath cell and of the three-carbon acid (pyruvate or alanine) returned to the mesophyll cell and (2) the nature of the enzyme that catalyzes the decarboxylation step in the bundle sheath cell. The three variants are named after the enzymes that catalyze the decarboxylation reactions. Representatives of each variant include maize, crabgrass, sugarcane, sorghum (NADP malic enzyme); pigweed, millet (NAD malic enzyme); guinea grass (phosphoenolpyruvate carboxykinase).

There are three variations of the basic C_4 pathway. The variations differ principally in the C_4 acid transported into the bundle sheath cells (malate or aspartate) and in the manner of decarboxylation.

The variations are named after the enzymes that catalyze the respective decarboxylation reactions: NADP-dependent malic enzyme (NADP-ME) found in chloroplasts; NAD-dependent malic enzyme (NAD-ME) in mitochondria; and phosphoenolpyruvate carboxykinase (PEP-CK) in the cytosol. The discussion in the text traces the route of carbon for the NADP-ME type, such as sugarcane or corn.

The Carbon Path

The primary carboxylation reaction, catalyzed by PEP carboxylase, which is common to all three variants, occurs in the cytosol of the mesophyll cells and uses HCO_3^- rather than CO_2 as a substrate. In the NADP-ME variant, oxaloacetate is rapidly reduced to malate in the mesophyll chloroplasts by NADPH using NADP:malate dehydrogenase. The malate formed enters the chloroplast of the bundle sheath cell and there undergoes oxidative decarboxylation, yielding pyruvate. The CO_2 released within the bundle sheath cells is converted to carbohydrate by the Calvin cycle.

The residual C_3 acid is transported back to the mesophyll as pyruvate and converted to phosphoenolpyruvate in the mesophyll chloroplast. This reaction takes place in the mesophyll chloroplast of all three C_4 variants. The cycle then starts again with carboxylation of the phosphoenolpyruvate to produce oxaloacetate.

Another process localized in the mesophyll chloroplast of all three C_4 variants is the reduction of 3-phosphoglycerate to triose phosphates, which occurs within the Calvin cycle proper in C_3 chloroplasts. In contrast, all C_4 variants export 3-phosphoglycerate from the bundle sheath for reduction in the mesophyll chloroplast. After reduction, the triose phosphates are transported back to the bundle sheath chloroplasts, for use in the operation of the Calvin cycle there.

One possible reason for this remarkable split of the Calvin cycle between the bundle sheath and the mesophyll chloroplast is to restrict electron transport activity and oxygen evolution in the bundle sheath, and thus avoid oxygen levels that would favour photorespiration. The transport of metabolites of the C_4 pathway is accompanied by movements of protons and other ionic species in order to maintain pH and charge balance.

Photorespiration in CAM Plants

Many terrestrial plants, adapted to hot and arid regions where water is scarce, are stressed by the severe effects of full sunlight, particularly drought and high temperature. To avoid high rates of water transpiration, the stomata of these species are tightly closed during the hot diurnal hours and perform most of their gas exchange with the surrounding atmosphere at night, when the air is relatively cool and humid. On the other hand, the vital process of photosynthesis in which CO_2 is incorporated into carbohydrates takes place during the day. To cope with the lack of an efficient diurnal influx of atmospheric CO_2, these species have evolved a different photosynthetic mechanism, called "crassulacean acid metabolism" (CAM).

This unusual process takes place within single leaf cells in which the uptake and subsequent assimilation of atmospheric CO_2 involves three compartments: cytosol, vacuole, and chloroplast. In general, CAM plants are common among desert succulents (e.g., *Cactaceae*) exposed to daytime high temperatures and low relative humidity, or among tropical forest epiphytes (*Bromeliaceae*) which are exposed to physiological drought under humid conditions because of their aerial growth habit. Remarkably, CAM also occurs in aquatic species (*Isoetaceae*) found worldwide in marshy and shallow habitats that exhibit a diurnal limitation of CO_2.

During the cool night-time hours, the open stomata facilitate the uptake of atmospheric CO_2 which diffuses into the cell cytosol, generating HCO_3^-. At this stage, phosphoenolpyruvate-carboxylase (PEPCase) catalyzes the incorporation of HCO_3^- into phosphoenolpyruvate, yielding oxaloacetic acid and related four-carbon organic acids (particularly malic acid), that accumulate in the vacuole.

During the hot hours of the day, the stomatal closure characteristic of terrestrial CAM plants reduces water loss but causes an internal CO_2 shortage. To overcome this deficiency, the vacuole releases malic acid into the cytosol where the action of decarboxylating enzymes, NADP–malic enzyme or PEP-carboxykinase, produces pyruvate and CO_2. Inside the stroma of chloroplasts, Rubisco incorporates the released CO_2 into 3-phosphoglycerate which, in concert with the ATP and NADPH synthesized in the light via photophosphorylation, initiates a series of Benson–Calvin cycle reactions that result in the accumulation of starch.

The temporal separation of the nocturnal formation and storage of malic acid, and the diurnal decarboxylation and the accompanying accumulation of plastid starch, proceeds over the course of a 24-hour photoperiod. On this temporal basis, CAM consists of four metabolic phases that encompass the modulation of C_4 (PEPCase) and C_3 (Rubisco) carboxylation within the same cellular environment (Figure 1). The nocturnal uptake of atmospheric CO_2 via PEPCase to form C_4 acids and the diurnal decarboxylation of malic acid to elicit an elevated internal concentration of CO_2 are interspersed with transitional periods of net CO_2 uptake in early morning and in late afternoon, when both the PEPCase-and the Rubisco-mediated carboxylations contribute to the assimilation of CO_2. In this circadian rhythm, the proportion of CO_2 taken up via PEPCase at night or via Rubisco during the day is controlled by:

- stomatal function
- fluctuations in the content of organic acids and storage carbohydrates
- the activity of enzymes involved in the primary (PEPCase) and the secondary (Rubisco) carboxylations, and in the decarboxylation of organic acids (e.g., malic enzyme or PEP-carboxykinase)

In diurnal phase III, the closed stomata preclude gas exchange with the external atmosphere, but photosynthesis and other vital processes continue. At this stage, the internal levels of CO_2 fall precipitously due to an active Benson–Calvin cycle; however, this decline is offset by mitochondrial respiration. The closure of the stomata also leads to the accumulation of O_2 in leaf cells because the rate of photosynthetic O_2 evolution is higher than mitochondrial respiration and other O_2-consuming reactions. At this stage, the oxygenase activity of Rubisco activates the photorespiratory cycle by which the Benson–Calvin cycle recovers 75% of the carbon diverted to glycolate, with the remaining 25% released as CO_2 by the mitochondrial glycine decarboxylase complex. Thus, the closed stomata provide an adaptive advantage that minimizes water loss from the leaves by transpiration, and keeps the concentration of stromal CO_2 adequate for carbon assimilation. However, the concurrent accumulation of O_2 is a major adverse consequence because, as a powerful oxidant, O_2 reacts with other molecules and generates harmful free radicals (e.g., superoxide, hydroxyl ion) which may initiate uncontrolled adverse reactions on DNA, proteins, and lipids.

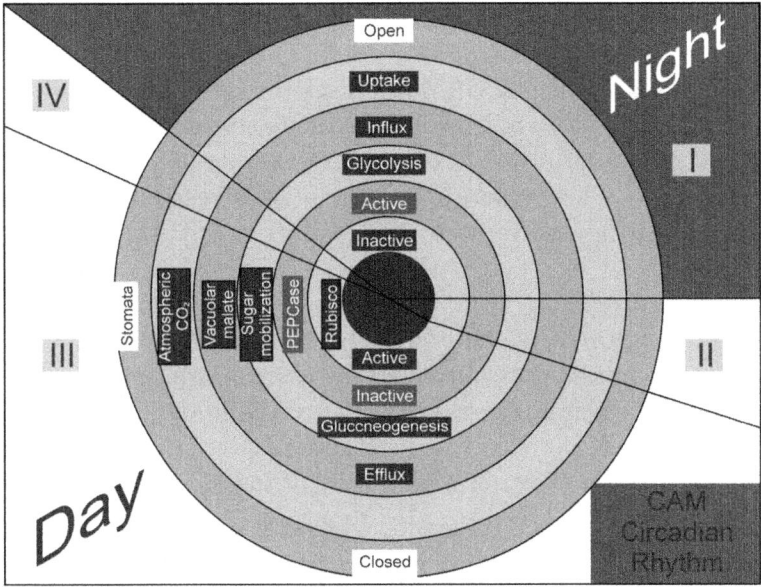

Figure : CAM Circadian rhythm. Metabolite fluxes [atmospheric CO₂, vacuolar malate, sugar mobilization to form PEP] and enzyme activities [PEPCase, Rubisco] are indicated on the four major temporal phases of CAM [I, II, III, IV]. Phase I comprises the nocturnal PEPCase-driven uptake of atmospheric CO_2 and malic acid accumulation. Phase II is the transitional phase wherein an accelerated burst of CO_2 uptake takes place due to both the PEPCase-mediated and the Rubisco-mediated fixation of CO_2. During diurnal Phase III, the uptake of atmospheric CO_2 is diminished and the decarboxylation of the vacuolar malic acid generates the stromal concentration of CO_2 needed for Rubisco-mediated CO_2 fixation via the Benson–Calvin cycle. Phase IV may include CO_2 uptake that is fixed directly by Rubisco

In this scenario, photorespiration helps alleviate the undesirable consequences of O_2 accumulation by consuming both stromal and peroxisomal O_2. The release of one mole CO_2 in a single round of the photorespiration cycle requires the successive utilization of: (a) two moles O_2 by the oxygenase activity of Rubisco in the chloroplast stroma, and (b) one mole O_2 for the concerted action of glycolate oxidase and catalase in the small oxygen cycle in the peroxisome. The photorespiratory cycle driven by ribulose-1,5-bisphosphate {2 Ribulose-1,5-bisphosphate + 3 O_2 + H_2O + ATP '! 3 3-phosphoglycerate + CO_2 + 2 P_i + ADP} requires substantially more O_2 than mitochondrial respiration operating with carbohydrates {$C_6H_{12}O_6$ (e.g., glucose) + 6

O_2 '! 6 CO_2 + 6 H_2O}. Thus, three versus one moles O_2 are consumed, respectively, per mole CO_2 released. It is therefore evident that the beneficial consequence of CAM photorespiration in Phase III involves not only the avoidance of carbon loss due to the formation of glycolate, but also the amelioration of oxidative stress due to removal of reactive oxygen species. This mitigation of oxidative stress has been proposed to be a major driving force in the evolution of CAM.

Fructans

Although starch is the dominant carbohydrate reserve in many species, it may be almost absent in the storage organs of some plants. Fructans are less familiar reserve carbohydrates but occur in 15% of flowering species that mainly belong to the Asteraceae, Campanulaceae and Boraginaceae (dicots), and Poaceae and Liliaceae (monocots).

This type of polysaccharide, whose basic unit is fructose, was found in 1804 in a hot water extract from *Inula helenium* and later named "inulin." Fructans can replace starch as a reserve carbohydrate in some plants, though in many species both polysaccharides coexist. Moreover, the carbohydrate reserve substances vary among plant species and even within a single plant species throughout the seasons and under various climatic conditions. In contrast to starch—which is stored in plastids—fructans appear in vacuoles of different organs: for example, in bulbs (onion), tubers (Jerusalem artichoke, dahlia), taproots (chicory), leaves, and in stems (wheat, barley).

Chemically, fructans are distinguished on the basis of the glycosidic bond that links fructose residues to each other. The biosynthesis of these polysaccharides starts with the incorporation of a fructose moeity to one of the three primary hydroxyl groups of vacuolar sucrose—that is, carbon-1 and carbon-6 in the fructose moeity, and carbon-6 in the glucose moeity. This reaction, catalyzed by the enzyme sucrose:sucrose 1-fructosyl transferase, implies a transglycosylation in which one molecule of sucrose transfers the fructosyl moeity to another. Thus, the addition of β-D-2,1-or β-D-2,6-linked fructofuranosyl units to the fructose moeity of sucrose yields 1-kestose and 6-kestose, respectively, while the addition of a β-D-2,6-linked fructofuranosyl unit to the glucose moeity of sucrose produces neo-kestose.

Further elongation of the ketose-type trisaccharide proceeds by the action of specific enzymes that transfer, also via transglycosylation, a fructosyl unit from a fructan molecule to positions 1 or 6 in the second fructose moeity of the trisaccharide. Thus, fructan:fructan fructosyl transferases catalyze the elongation of kestoses yielding the linear β-D-2,1-bonds of inulins, mainly in dicots, and the β-D-2,6-linkages in levans of monocots and bacteria.

$$GFF \text{ (kestose)} + G(F)_m \text{ '!} G(F)_3 + G(F)_{m-1}$$

While up to 100,000 fructose moeities can be linked in a single molecule of bacterial fructan, fructosyl units range from 10 to 200 in plant counterparts. Fructans generally are present as a continuum of oligofructans that differ from each other in one fructose residue: $G(F)_3$, $G(F)_4$, $G(F)_5$, $G(F)_6$,..., $G(F)_n$. This feature enables fructans to participate in the regulation of the cellular osmotic potential.

In addition to supplying hexose units for demands of cellular energy, the breakdown of the endogenous polysaccharide elevates the concentration of lower members in the oligofructans series, quickly increasing in consequence the osmotic pressure.

Two types of fructan hydrolases cleave the glycosidic bond in fructans: exohydrolases that release terminal fructosyl residues while endohydrolases cleave at random within the fructan chains. Apparently, plants contain only the former type while bacteria and fungi contain both kinds of enzymes.

The recent finding of fructan exohydrolases in nonfructan plants not only suggests that these enzymes are additionally involved in other important functions—perhaps defence and signalling—but also highlights the necessity of functional characterization of these enzymes (Van den Ende et al. 2004).

Plastidic Phosphate Translocators

Chloroplasts are metabolic powerhouses within leaf cells. They are not only the main site of carbon, nitrogen, and sulfur assimilation but also share metabolic pathways—such as the photorespiratory carbon cycle—with other plant cell compartments. A continuous exchange of metabolites and ions with the cytosol is necessary for playing these roles.

In all vascular plants, plastids are surrounded by an envelope that restricts the nonspecific diffusion of polar molecules. Thus, a pair of concentric membranes modulate the exchange of metabolites between the stromal interior and the cytosol. Although many pore-forming proteins that are substrate-specific were characterized in the outer membrane, the inner membrane is considered the main permeability barrier between the cytosol and the chloroplast. All transporters of the inner membrane are nuclear-encoded membrane proteins whose precursors are synthesized in the cytosol and post-translationally imported into the inner membrane.

Chloroplast transporters collectively grouped as phosphate translocators catalyze a strict 1:1 counter-exchange (antiport) of orthophosphate with triose phosphates, pentose phosphates, and hexose phosphates. The Arabidopsis genome contains six genes that encode four classes of functional plastidial phosphate translocators: single copies of the triose phosphate/phosphate translocator (TPT) and the phosphoenolpyruvate/phosphate translocators (PPT); and two copies each of the glucose 6-phosphate/phosphate translocators (GPT) and the xylulose 5-phosphate/phosphate translocator (XPT).

Table : Plastidic Phosphate Translocators

Trans-locator	MATDB entry	Compounds transported
TPT	TPT (At5g46110)	triose phosphates, 3-phosphoglycerate
PPT	PPT1 (At5g54800), PPT2 (At3g01550)	phosphoenolpyruvate, 2-phosphoglycerate
GPT	GPT1 (At5g33320), GPT2 (At1g61800)	glucose 6-phosphate, triose phosphates
XPT	(At5g17630)	xylulose 5-phosphate, triose phosphates

The Triose Phosphate/Phosphate Translocator and the Allocation of Carbon in Plants

Triose phosphates, generated by the Calvin cycle at the expense of photosynthetically generated ATP and NADPH, flow to the cytosol across the chloroplast envelope. In turn, the orthophosphate released in the cytosol from biosynthetic processes is transported back into chloroplasts to replenish the ATP necessary for sustaining the assimilation of CO_2 and photosynthetic electron transport. Both the efflux of triose phosphates and the influx of orthophosphate are driven by a dimer composed of two identical subunits—the chloroplastic TPT. This membrane protein functions as an antiport system that exchanges phosphorylated—mainly three-, but also admits five-or six-carbon compounds—for orthophosphate.

Hence, the chloroplast TPT exports the fixed carbon for the cytosolic synthesis of sucrose, the main product allocated to heterotrophic plant organs for further metabolism or conversion into storage products (e.g., starch, fructans). Therefore, the chloroplast TPT represents *the diurnal path for carbon export*.

Mutants defective in chloroplast TPT rely on different paths to transfer the photoassimilate from the chloroplast to the cytosol and, in so doing, compensate for the deficiency in TPT activity.

Therefore, knockout mutations of this translocator increase the allocation of recently fixed carbon into transitory starch, followed by the degradation of the starch in the light and the subsequent export of neutral sugars to the cytosol. On the other hand, the simultaneous knockout of chloroplast TPT and starch synthesis drastically impairs plant growth.

The Phosphoenolpyruvate/Phosphate Translocator is Essential for the Chloroplast Metabolism

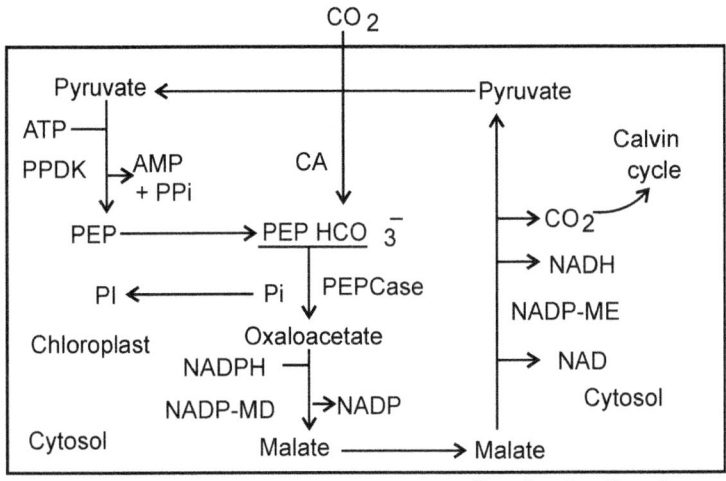

Figure : Mobilization of phosphoenolpyruvate in mesophyll chloroplasts of C_4 plants of the NAD-malic enzyme type. The pyruvate produced by the malic enzyme (ME) in bundle sheath cells flows to mesophyll cells for conversion to phosphoenolpyruvate (PEP) by the pyruvate-phosphate dikinase (PPDK) localized at the chloroplast stroma. The efflux of PEP to the cytosol through the phosphoenolpyruvate/phosphate translocator (PPT) restores the acceptor of HCO_3 required by the phosphoenolpyruvate carboxylase (PEPCase) for the primary carbon fixation. CA: carbonic anhydrase; NADP-MD: NADP-dependent malate dehydrogenase.

PPT in the inner chloroplast membrane also links the stromal metabolism of all plants with the surrounding cytosol. C_4-plants of the NADP-malic enzyme type (e.g., maize) under active photosynthesis produce pyruvate in bundle sheath cells. This three-carbon compound returns to mesophyll cells for the conversion to phosphoenolpyruvate by the chloroplast-localized pyruvate-phosphate dikinase (Figure below). Subsequently, the phosphoenolpyruvate flows from the stroma to the cytosol for the primary carbon fixation via PEPCase. To ensure high rates of phosphoenolpyruvate efflux for the proper functioning of this cycle, the envelope of mesophyll chloroplasts bears the PPT, which concurrently drives the chloroplast uptake of orthophosphate to restore the lost phosphorus.

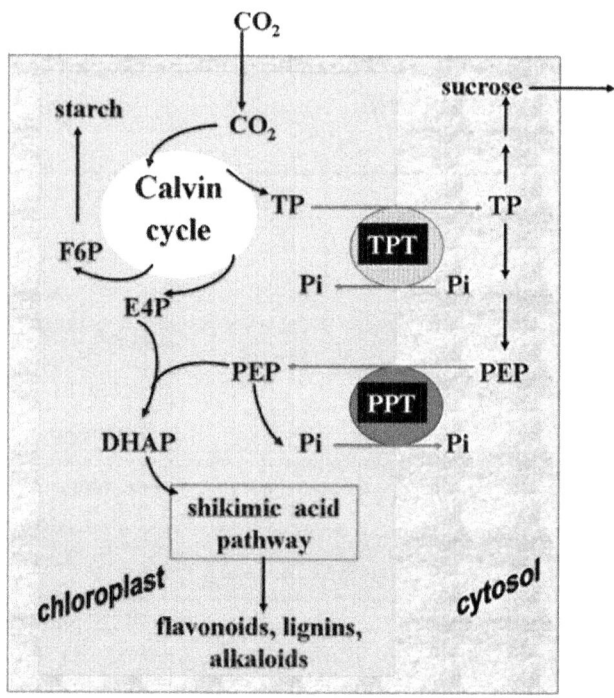

Figure : Mobilization of phosphoenolpyruvate in chloroplasts of C₃ plants. Under active photosynthesis, TPT furnish the cytosol with carbon skeletons for the biosynthesis of phosphoenolpyruvate. Upon the release of orthophosphate, the uptake of the cytosolic phosphoenolpyruvate, which combines in the stroma with erythrose 4-phosphate, triggers the formation of 3-deoxy-D-arabino-heptulosonic acid-7-phosphate (DHAP). DHAP is a scaffold for the biosynthesis of shikimic acid-5-phosphate, which in turn starts the pathway leading to the aromatic amino acids tyrosine and phenylalanine, the precursors of flavonoids, lignins, and alkaloids.

On the other hand, CAM plants of the malic enzyme-type decarboxylate during the day the malate accumulated at night. Again, two successive processes account for the production of phosphoenolpyruvate, the primary acceptor of CO_2. First, the pyruvate formed in decarboxylation of the malate enters into the chloroplast stroma for its transformation to phosphoenolpyruvate catalyzed by the pyruvate-phosphate dikinase. Second, the phosphoenolpyruvate leaves the chloroplast in exchange for orthophosphate via the PPT.

Knowledge on the function of the PPT in C_3 plants or nongreen tissues of C_4 plants is scarce. Apparently, the uptake—rather than the export—of phosphoenolpyruvate is the primary function of PPT in these tissues. Chloroplasts and most nongreen plastids are devoid of phosphoglyceromutase and enolase and, consequently, are unable to convert 3-phosphoglycerate to phosphoenolpyruvate. However, phosphoenolpyruvate combines with erythrose 4-phosphate, an intermediate of the pentose-phosphate pathway, for the synthesis of aromatic compounds through the shikimic acid pathway (Figure below). Thus, the provision of cytosolic phosphoenolpyruvate to the stroma via the PPT would be essential for the biosynthesis of flavonoids, lignins, alkaloids.

The Glucose 6-Phosphate/Phosphate Translocator Functions in Nongreen Tissues and is Essential for Plant Development

The genome of Arabidopsis harbours two paralogous GPT genes, *AtGPT1* and *AtGPT2*, that, apparently, function in plastids of nongreen tissues for the import of glucose 6-phosphate. Two transgenic lines, bearing T-DNA insertions in the *GPT1* gene, exhibited not only impaired pollen and ovule development but also were lethal in the homozygous state. At variance, the disruption of the *GPT2* genes have no effect on growth and development. On the basis of drastic gametogenesis defects, it appears that the import of glucose 6-phosphate into nongreen plastids by the translocator AtGPT1 is crucial for the maturation of the pollen and the development of the female gametophyte.

The Xylulose 5-Phosphate/Phosphate Translocator

The XPT constitutes the fourth subfamily of phosphate translocators that exhibits substrate specificity similar to the GPT but lacks the capacity to transport glucose 6-phosphate. In Arabidopsis, the lack of cytosolic transketolase and transaldolase halts the cytosolic oxidative pentose phosphate pathway at the stage of pentose phosphates (Kruger and von Schaewen 2003). In this context, the putative function of the XPT would be to furnish the stromal pentose phosphate pathway with cytosolic five-carbon skeletons in the form of xylulose 5-phosphate, especially under conditions that require its removal from the cytosol.

Photosynthesis: Physiological and Ecological Considerations

Working with Light

Three light properties are especially important when working with light: amount, direction, and spectral quality. The first two parameters, amount and direction, are important with respect to the geometry of the part of the plant that intercepts the light. Is the plant part flat or cylindrical? For a flat leaf, a planar light sensor is the most appropriate, and the amount of energy that falls on a flat sensor of known area per unit time is quantified as irradiance. Units can be expressed in terms of energy, such as watts per square meter (W m^{-2}). Time (seconds) is contained within the term watt: 1 W = 1 joule (J)s^{-1}. The energy of a photon depends on its frequency, as expressed by Planck's law.

When considered as a wave, light has a wavelength and a frequency. Light can also be thought of as a stream of particles, photons, or quanta. In this case, units can be expressed in moles per square meter per second (mol m^{-2} s^{-1}), where "moles" refers to the number of photons (1 mol of light = 6.02×10^{23} photons, Avogadro's number). This measure is called photon irradiance.

Quanta and energy units can be interconverted, provided that the wavelength of the light is known. The energy of a photon is related to its wavelength as follows:

$$E = \frac{hc}{\lambda}$$

where c is the speed of light (3×10^8 m s^{-1}), h is Planck's constant (6.63×10^{-34} J s), and λ is the wavelength of light, usually expressed in   nm (1 nm = 10^{-9} m). We can solve for the $h\ddot{e}$ part of the equation, and we obtain $1,988 \times 10^{-16}$, and write this equation as:

$$E = \frac{1,988 \times 10^{-16} \, J}{\lambda}$$

where λ is expressed in nanometers. From this equation we can see that a photon at 400 nm, which is in the blue region of the spectrum, has twice the energy of a photon at 800 nm, from the infrared region of the spectrum. A photon of 400 nm light contains 4.97×10^{-19} J. On the other hand, the 800 nm photon contains

2.48 × 10^{-19} J. Stated differently, the higher the wavelength of a photon, the lower its energy, as indicated by the larger denominator in the equation.

Now, suppose we have 3 µmol of 400 nm light falling on 1 m^2 every second, or a photon irradiance of 3 µmol m^{-2} s^{-1}. This quantity is approximately the amount of blue light at 400 nm that strikes the surface of Earth on a sunny day. If we want to convert photon irradiance to energy irradiance, we must first convert micromoles to moles (1 µmol = 10^{-6} mol): (3 µmol m^{-2} s^{-1}) × (10^{-6} mol µmol^{-1}) = 3 × 10^{-6} mol m^{-2} s^{-1} of photons, or quanta. Calculating the number of quanta from Avogadro's number gives: (3 × 10^{-6} mol quanta m^{-2} s^{-1}) × (6.02 × 10^{23} quanta mol^{-1}) = 1.8 × 10^{18} quanta m^{-2} s^{-1}. As we already calculated, each photon, or quantum, of 400 nm light contains 4.97 × 10^{-19} J. Thus (4.97 × 10^{-19} J quantum^{-1}) × (1.8 × 10^{18} quanta m^{-2} s^{-1}) = 0.9 J s^{-1} m^{-2}. Since 1 J s^{-1} = 1 W, we have an irradiance of 0.9 W m^{-2}.

Direction of light. Turning our attention to the direction of light, light can strike a flat surface directly from above or it can strike the surface obliquely. When light deviates from perpendicular, irradiance is proportional to the cosine of the angle at which the light rays hit the sensor.

Thus, irradiance is maximal when light strikes a surface directly from above, and it decreases as light becomes more oblique—similar to the situation with a typical leaf. Sensors that correct for the angle of incidence of light are said to be *cosine corrected*.

There are many examples in nature in which the light-intercepting object is not flat (e.g., complex shoots, whole plants, chloroplasts). In addition, in some situations light can come from many directions simultaneously (e.g., direct light from the sun plus the light that is reflected upward from sand, soil, or snow). In these situations it makes more sense to measure light with a spherical sensor that measures light omnidirectionally (from all directions).

When the amount of light is measured by this omnidirectional measurement, the type of measurement is called fluence rate (Rupert and Letarjet 1978), and the measured amount of light can be expressed in watts per square meter (W m^{-2}) or moles per square meter per second (mol m^{-2} s^{-1}). It is clear from the units whether light is being measured as energy (W) or as photons (mol).

In contrast to a flat sensor, a spherical sensor is equally sensitive to light from all directions. Depending on whether the light is collimated (rays are parallel) or diffuse (rays travel in random directions), values for fluence rate versus irradiance can be quite different from one another. They are equivalent only under special conditions.

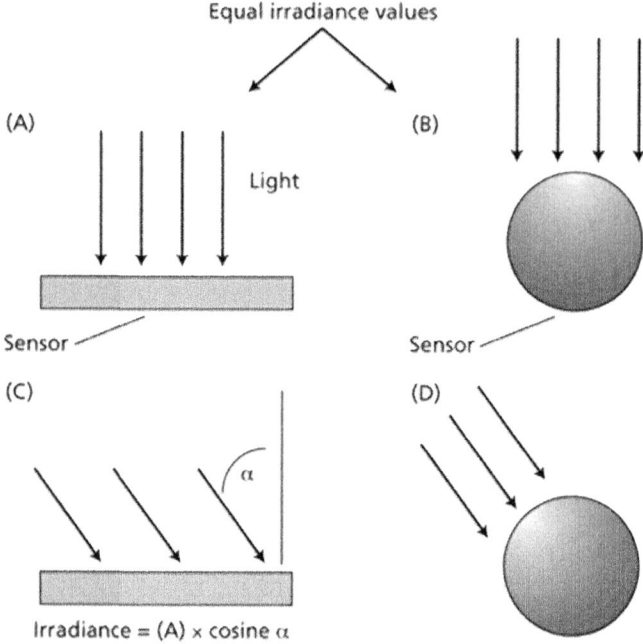

Figure: Irradiance and fluence rate. Equivalent amounts of collimated light strike a flat irradiance-type sensor (A) and a spherical sensor (B) that measure fluence rate. With collimated light, A and B will give the same light readings. When the light direction is changed 45°, the spherical sensor (D) will measure the same quantity as in B. In contrast, the flat irradiance sensor (C) will measure an amount equivalent to the irradiance in A multiplied by cosine of the angle α in C. (After Björn and Vogelmann 1994.)

The flat sensor measurement of photosynthetically active radiation (PAR, 400 to 700 nm) may also be expressed on the basis of energy (W m^{-2}) or quanta (mol m^{-2} s^{-1}) (McCree 1981). It is important to note that PAR is an irradiance-type measurement. In research on photosynthesis, when PAR is expressed on a quantum basis, it is often given the special term photosynthetic photon flux density (PPFD). However, it has been suggested that the term "density" be discontinued because within the International System of Units (Système Internationale d'Unités, or SI units) "density" can mean area or volume. Moreover, area is contained within the term flux. PPFD has in some cases been shortened to PPF, but it is not clear whether this abbreviation represents an irradiance-type or a spherical measurement.

When light moves through water such as lakes or oceans, it scatters in such a way that its *direction* of travel is randomized. As a result, aquatic photosynthetic organisms can receive light simultaneously from above, below, and the sides. In oceanography and limnology, the light that travels from the water surface toward the bottom is called *downwelling irradiance*, and the light that travels in the reverse direction is called *upwelling irradiance*.

Before the advent of specialized light meters suitable for plant physiology, light was measured in lux or foot-candles. These measurements are based on the perception of light by the human eye, which is maximally sensitive to light within the green region of the spectrum, at 555 nm. Sensitivity falls off on both sides of this wavelength and approaches zero within the blue and red, wavelength regions that are important for photosynthesis. Thus, lux and foot-candles are of little use in plant physiology. These units can be converted to the units of irradiance and fluence rate, but only if detailed knowledge about the spectral quality of the light is available, something that can be obtained only by use of an instrument called a *spectroradiometer*, which measures the amount of light at each wavelength.

In summary, when choosing how to quantify light, it is important to match sensor geometry and spectral response with that of the plant. Flat, cosine-corrected sensors are ideally suited to measure the amount of light that strikes the surface of a leaf; spherical sensors are more appropriate in other situations, such as when studying a chloroplast suspension or a branch from a tree.

How much light is there on a sunny day and what is the relationship between PAR irradiance and PAR fluence rate? Under direct sunlight, PAR irradiance and fluence rate are both about 2000 μmol m^{-2} s^{-1}, though higher values can be measured at high altitudes. The corresponding value in energy units is about 400 W m^{-2}. When light is completely diffuse, irradiance is only 0.25 times the fluence rate.

Heat Dissipation from Leaves: The Bowen Ratio

The heat load on a leaf exposed to full sunlight is very high. In fact, a leaf with an effective thickness of water of 300 μm would warm up by 100°C every minute if all available solar energy were absorbed and no heat was lost. However, this enormous heat load is dissipated by the emission of long-wave radiation, by sensible (or perceptible) heat loss, and by evaporative (or latent) heat loss.

Air circulation around the leaf removes heat from the leaf surfaces if the temperature of the leaf is higher than that of the air; this phenomenon is called sensible heat loss.

Evaporative heat loss occurs because the evaporation of water requires energy. Thus, as water evaporates from a leaf it withdraws heat from the leaf and cools it. The human body is cooled by the same principle, through perspiration.

Sensible heat loss and evaporative heat loss are the most important processes in the regulation of leaf temperature, and the ratio of the two is called the Bowen ratio:

$$\text{Bowen ratio} = \frac{\text{sensible heat loss}}{\text{evaporative heat loss}}$$

This concept was developed by Ira S. Bowen (1898–1978), an American astrophysicist. When the evaporation rate is low, because water supply is limited, the Bowen ratio tends to be high. Thus, the Bowen ratio is about 10 for deserts, 2-6 for semi-arid regions, 0.4 to 0.8 for temperate forests and grasslands, 0.2 for tropical rain forests and 0.1 for tropical oceans

In well-watered crops, transpiration, and hence water evaporation from the leaf, is high, so the Bowen ratio is low. On the other hand, in some cacti, stomata closure prevents evaporative cooling; all the heat is dissipated by sensible heat loss, and the Bowen ratio is infinite. Plants with very high Bowen ratios conserve water but have to endure very high leaf temperatures in order to maintain a sufficient temperature gradient between the leaf and the air. Slow growth is usually correlated with these adaptations.

Very low Bowen ratios can be measured in a lawn on a relatively still day. In these conditions there is no sensible heat loss, because the air around the leaf is at the same temperature as the leaf; the Bowen ratio therefore approaches zero, and heat dissipation is due mostly to evaporative heat loss.

In other cases, such as cotton leaves in the afternoon, water loss from stomatal transpiration cools the leaf below the air temperature by evaporative heat loss. In that case, there is sensible heat gain rather than heat loss, and the Bowen ratio becomes negative.

One can calculate the evapotranspiration rate for an entire canopy using measurements of the Bowen ratio, net incident radiation, the heat loss from the soil, and the gradients in temperature and water vapour concentration above the canopy.

The Geographic Distributions of C_3 and C_4 Plants

Among the 15,000+ species with C_4 photosynthesis, it is most common in grasses and sedges, less common in herbs and shrubs, and not found in trees (with a single Hawaiian tree exception, *Euphorbia forbesii*). Climate is a major factor influencing the natural distributions of C_3 plants and C_4 grasses. Here, we talk about the two most important climate parameters influencing plant growth: water and temperature. Clearly plants will not grow in the absence of water, so the important factor influencing photosynthetic pathway distribution becomes the temperature during the growing period. Based on many systematic surveys of the natural vegetation across the globe that have been accumulating over time, a clear picture is emerging. C_4 taxa are found in warm to temperate environments and are uncommon in cool to cold climates. Below we construct a global map that describes the general abundances of C_3 and C_4 taxa on different continents.

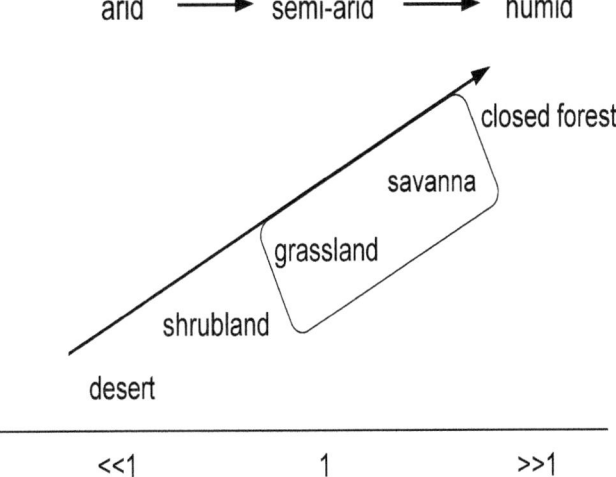

Figure: A plot of the vegetation types that are expected to occur along a climate gradient estimated by the ratio of the incoming precipitation to the potential evaporation rate. Courtesy of Ehleringer, unpublished figure.

Notice that C_4 taxa are not very common in tropical regions (0–20° latitude), because dense tropical forests tend to shade out C_4 grasses. The C_4 taxa are most common away from the tropics in the savanna and steppe regions; their abundances tends to diminish south of the desert zones (generally 30–40° latitude). Climate can be presented along an axis of an increasing of precipitation relative to evaporative water loss. At the upper, wetter end of this gradient, we would find the forests. At the drier end, where precipitation is low relative to evaporative demand, we see the deserts. In the figure below, we see that these different vegetation zones reflect variations in the ratios of precipitation-to-evaporation. C_4 plants are most common in this semi-arid region.

In deserts with wet monsoon summers, such as in central Africa, the sub-Indian continent, and western North America, C_4 taxa are abundant. In deserts that receive winter rainfall, such as deserts in Asia and North Africa, there are few C_4 taxa. In North America, this regional influence of summer precipitation is most visible in the southwestern United States where there are large differences in precipitation associated with the Arizona monsoon boundary as shown below.

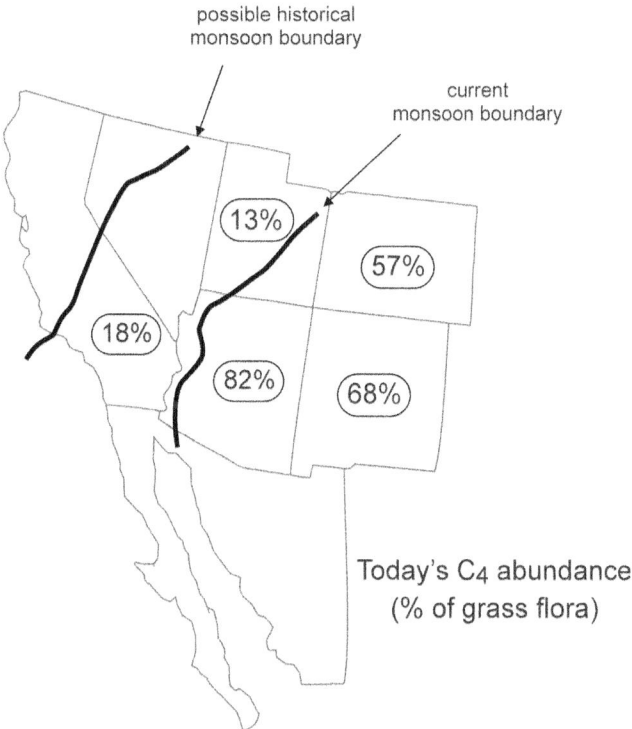

Figure: Calculations of the percentage of the grass species that have C_4 photosynthesis in different states of the western United States. Note that the presences of summer rain influences C_4 abundance. C_4 plant tend to occur in warm, summer wet regions. Courtesty of Ehleringer, Cerling, and Dearing (2005).

Human activities and disturbances by animals will influence the distribution of C_3 and C_4 taxa within savannas and grasslands. This reflects the role of disturbances, such as grazing and fires, on the importance of trees on the landscape.

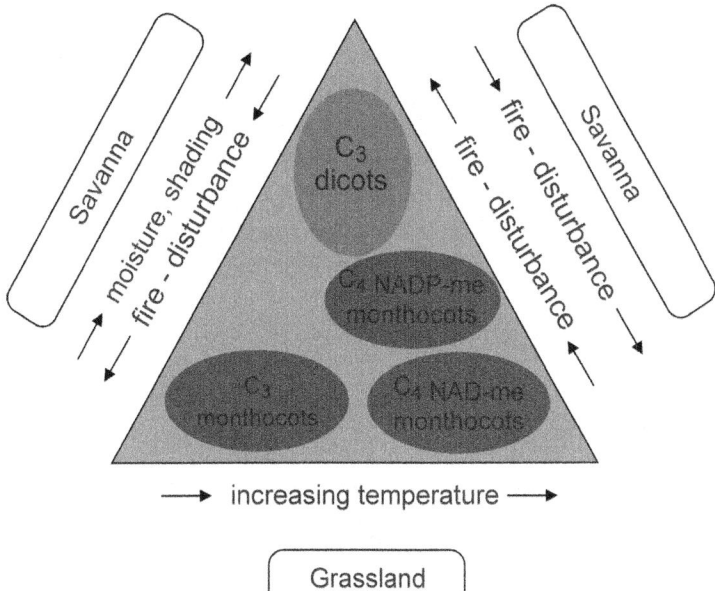

Figure: A three-axis, triangular presentation of the different gradients influencing the abundances of different C_4 photosynthesis subtypes. Courtesy of Ehleringer, Cerling, and Dearing (2005).

Across the grasslands and savannas, the two dominant C_4-photosynthesis subtypes do not share identical distributions. The C_4 NADP-me grasses tend to occur in drier regions, such as the shortgrass prairie of the Great Plains. On the eastern, wetter edge of the Great Plains, the shortgrass prairie is replaced by tallgrass prairie, dominated by C_4 NAD-me grasses. Today, agricultural practices result in C_4 plants growing outside of the distributions shown above. This results from the extensive planting of corn, sugarcane, and millet in both tropical and temperate regions.

Projected Future Increases in Atmospheric CO_2

Human use of fossil fuels (coal, oil, and natural gas) continues to increase as growing human populations demand more energy for transportation, heating, and manufacturing. We measure atmospheric CO_2 in units of ppm or parts per million. The rate of atmospheric increase in CO_2 is about 3 ppm per year.

Note the cyclic nature of the atmospheric CO_2 data, in which one oscillation cycle is exactly one year. This annual pattern reflects changes in the balance of photosynthesis (decreases atmospheric CO_2) and respiration (increases atmospheric CO_2) at a location over the course of the year. Atmospheric CO_2 tends to decrease in the spring and summer, when photosynthesis rates within an ecosystem exceed respiration rates. In contrast, atmospheric CO_2 tends to increase in the fall and winter when respiration rates exceed photosynthesis. In 2007 atmospheric CO_2 reached an average value of 384 ppm and is expected to reach 400 ppm before 2015.

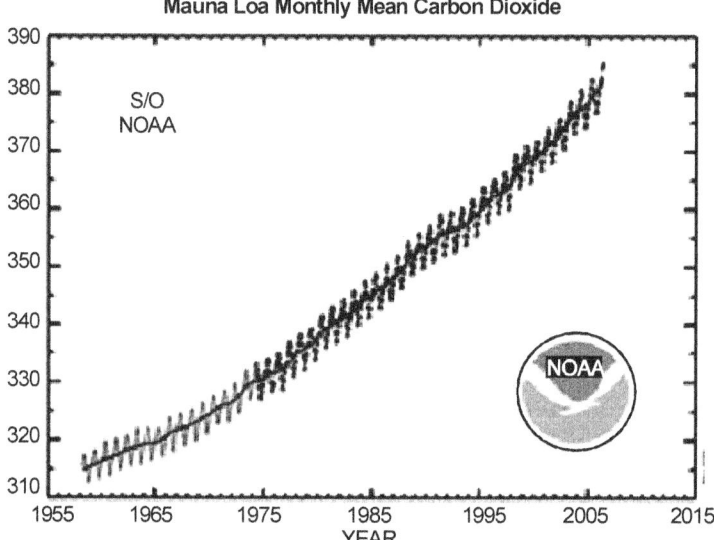

Figure: A high precision record of the atmospheric carbon dioxide levels measured on Manua Loa, Hawaii.

Economists have good estimates of the rate of CO_2 emission globally. The U.S., European countries, China, Japan, and India are the largest sources of fossil fuel emissions.

One surprising fact is that the observed rate of atmospheric CO_2 increase is actually less than the observed rate of atmospheric CO_2 increase. This is because plants on land and algae in the ocean are currently able to take up about one-half of fossil fuel emissions through enhanced photosynthesis. Scientists study how plants and ecosystems respond to elevated CO_2 using an experimental field approach called a Free Air CO_2 Enrichment (FACE) Experiment. In a FACE experiment, pipes inject CO_2 into the interior of a ringed area containing a complete ecosystem as shown below.

These FACE research facilities give scientists an opportunity to understand how different plant biochemical, physiological, and growth processes within the ecosystem will respond as a result of long-term exposure to elevated CO_2 levels.

Since biomass production involves so much more than simply increased photosynthesis (i.e., mineral nutrients are required as well), it is doubtful that plant growth can be sustained in a linear, proportional fashion as atmospheric CO_2 levels continue to increase.

The FACE studies are designed to address the question of how ecosystems will respond to future atmospheric CO_2 environments and whether the growth response level off at some future CO_2 level.

Global warming and changes in climate are anticipated effects of a rapidly increasing CO_2 levels. These are, of course, but two of the many reasons why scientists and others are concerned about the consequences of elevated atmospheric CO_2. Just how much the atmospheric CO_2 will increase is unknown. Below are estimates of the ranges, based on two plausible scenarios.

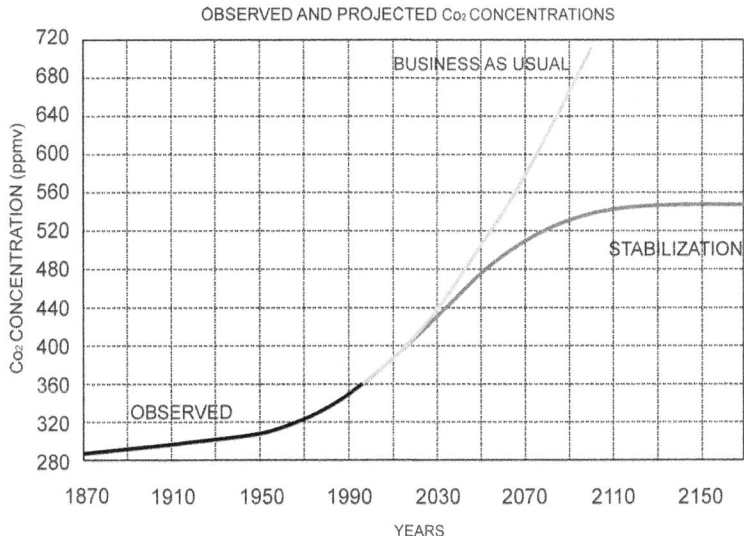

Figure: Projected changes in the atmospheric CO_2 concentrations under contrasting emission-control projections.

In one scenario, titled "business as usual" atmospheric CO_2 levels are projected to reach 700 ppm by the end of this century. On the other hand, an aggressive global effort to curb CO_2 emission might result in an atmospheric CO_2 stabilization of 550 ppm.

Using Carbon Isotopes to Detect Adulteration in Foods

Carbon isotope ratios ($\delta^{13}C$) of C_4 plants such as corn and sugar cane are -10 to -12 ‰, in contrast to $\delta^{13}C$ values of C_3 foods, which are typically -26 to -28‰

The naturally occurring variations in carbon isotope ratios within the C_3 of C_4 ranges occur because environmental factors influence carbon isotope ratios. However, the differences in carbon isotope ratios among the two photosynthetic pathways are large enough to ensure that it is possible to distinguish between whether or not a food substance is based on C_3 of C_4 photosynthesis.

Adulteration is the substitution of a cheaper food substance or compound with another. Carbon isotopes are frequently used to detect the substitution of C_4 sugars for C_3 food products, such as the introduction of sugar cane into honey to increase yield. One easy way to detect this adulteration is to examine the jams and jellies that are found in restaurants. Below are the results of a survey conducted as a science project by Jeff Ehleringer when he was a young man in junior high school. Jeff collected jams and jellies every time he went into a restaurant. He had the opportunity to collect jams and jellies from across the western United States and in parts of Europe as well.

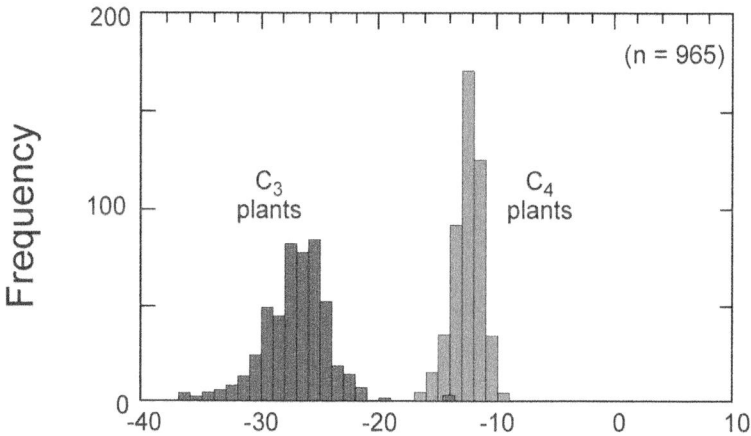

Carbon isotope ratio, ‰

Figure: Variations in the carbon isotope ratios of C_3 and C_4 plants. Courtesy of Cerling et al. (1997) Nature.

The results shown above are interesting and indicate that adulteration has taken place in some of the samples. The fruits that go into producing jams and jellies are all C_3 fruits (grapes, strawberries, raspberries, orange, apricots, peaches, etc.) One subgroup of the jams and jellies appear to have carbon isotope ratios similar to the stable isotope ratios of C_3 plants, from which they were supposed to have been derived. These were all samples that Jeff had collected in Europe, where there are authenticity laws regulating what materials can go into making a processed food such as a jelly. In contrast, the vast majority of the jams and jellies had carbon isotope ratios consistent with those of C_4 plants. Yet, fruits are not from C_4 plants but instead are produced from C_3 plants. These are the adulterated jellies, most likely produced largely from sugar syrup and food flavouring with very little food used in the process. These substitutions are legal in the U.S., because here, we do not have authenticity laws with respect to food production. A third small subgroup consists of jams that were made of 50/50 C_3/C_4 food sources. It turns out that these were the more expensive jams and jellies. There are many other food products that are tested for their carbon isotope ratios to detect adulteration. Common among these are fruit juices, vanilla (vanilla beans come from CAM plants), beer (wheat and barley are C_3), and alcoholic beverages.

Reconstruction of the Expansion of C₄ Taxa

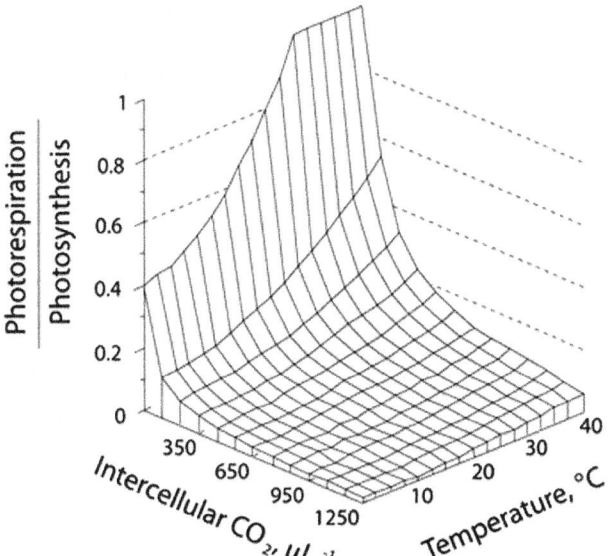

Figure: Changes in the ratio of photorespiration-to-photosynthesis as a function of both CO_2 concentration and temperature for a plant with C_3 photosynthesis. Courtesy of Ehleringer and Cerling (2001).

C$_4$ photosynthesis is favoured over C$_3$ photosynthesis under conditions of high temperature and/or low atmospheric CO$_2$. Below we provide a 3-D graphic of how the ratio of photorespiration-to-photosynthesis increases as temperature increases and/or CO$_2$ decreases.

The rate of photorespiration in a leaf increases because RuBP carboxylase is more likely to react with O$_2$ instead of CO$_2$ as temperature increases and/or CO$_2$ decreases. The result of an increase in photorespiration is a decrease in photosynthetic quantum efficiency. This leads to environmental conditions where C$_4$ plants are favoured over C$_3$ and vice versa.

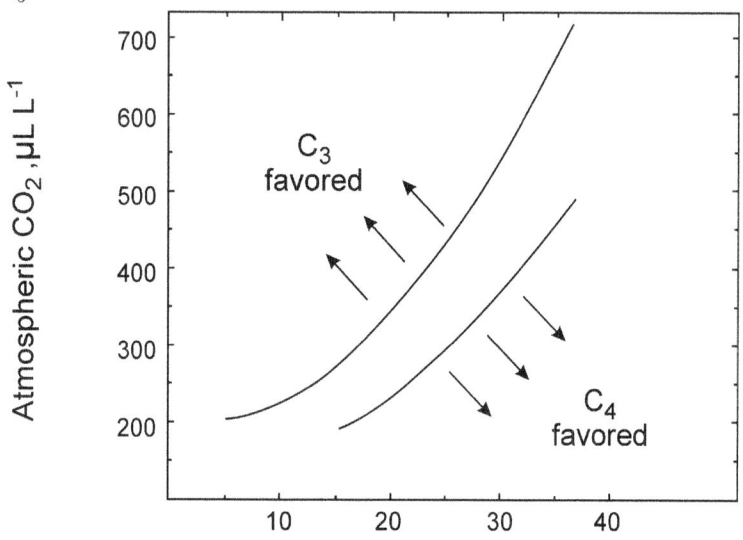

Daytime growing-season temperature, C

Figure: Predicted realms of atmospheric CO$_2$ and growing season temperature that favour plants with C$_3$ or C$_4$ photosynthesis.

This model leads to predictions that the abundances of C$_3$ and C$_4$ plants should vary spatially and as a function of atmospheric CO$_2$ history. The secret to evaluating this model is to go back and look at plant fossils over time and to relate photosynthesis pathway with prehistoric atmospheric CO$_2$. This is a difficult task.

Reconstruction of the expansion of C$_4$ plants over the past 10–15 million years can be a challenge, because individual plants are not well recorded in the fossil record in many locations. Most plant fossils are thought to have formed as a result of submersion into an anaerobic, aquatic environment. The vast majority of plants do not become fossils under these conditions, but instead are decomposed by microbial activities leading to diminished numbers of fossils that we can use to reconstruct the expansion of C_4 plants over time. However, there are good proxies.

Carbon isotope ratios ($\delta^{13}C$) of animal tissues reflect the foods that they ate. For modern animals, it is possible to choose from among, hair, muscle, bone collagen, and tooth enamel for isotope analyses in order to determine the proportions of C_3 and C_4 food sources in their diets. Hair provides a sequential record of the animal's diet during the time of hair production. Muscle and bone collagen analyses provide an integrated measure of the animal's current diet as shown in the figure below.

The $\delta^{13}13C$ of animal teeth faithfully record the carbon isotope ratios of food sources during the period of tooth development. The carbon in tooth enamel is actually carbonate that has condensed from CO_2 that was produced as a byproduct of metabolizing food. After an animal has died and decomposition has occurred, about the only remaining tissue that remains and preserves the C_3/C_4 dietary signals is the enamel in the animal's teeth.

For many herbivores, such as cows and horses, tooth growth is continuous and provides a longer-term dietary record than for animals with a fixed tooth growth period (such as humans). As teeth are preserved for millions of years, δ^{13} of tooth enamel can be used to reconstruct the abundances of C_3 and C_4 plants eaten by mammalian grazers over extended time periods. In the graphic below, we plot the frequency distributions of modern animal teeth with that observed for C_3 and C_4 plants:

Note that there is an "å" offset of 14.1‰. This is the isotope enrichment associated with CO_2 condensing to form carbonate. Now we have a tool—the carbon isotope ratio of animal tooth enamel will record the abundance of C_3 versus C_4 food sources.

Until 8 million years ago, C_4 plants were not an important part of ecosystems. Then, between 6 and 9 million years ago, C_4 plants became important parts of many ecosystems, particularly at tropical and semi-tropical latitudes. These results are consistent with model predictions above. C_4 photosynthesis is predicted not to occur on Earth until the atmospheric CO_2 level falls below a critical threshold value. Even then it should appear first in those locations with the warmest temperatures during the growing season.

There is also extensive evidence presently to show that fluctuations in the atmospheric CO_2 concentration values between glacial (180 ppm) and interglacial (280 ppm) periods were large to influence the local abundances of C_3/C_4 plants. Shown below are the time series analyses of carbon in bog in tropical Africa. At about 10,0000 to 12,000 years ago, the vegetation surrounding these bogs switched from being dominated by C_4 plants to being dominated by C_3 plants. This shift is evident by the dramatic change is carbon isotope ratios of materials feeding into the bog, especially at the time that the glacial period ended and we entered our current inter-glacial period.

Together these historical pieces of information paint an interesting history. The C_3 photosynthetic pathway dominated for much of Earth's history. And it is only relatively recently in Earth's (6–8 million years ago) history that we have seen an expansion of C_4-dominated ecosystems. What the future holds is unclear, since anthropogenic burning of fossil fuels is rapidly increasing in atmospheric CO_2 to levels far exceeding those observed on Earth over the past 1 million years.

8

Respiration and Lipid Metabolism

Although it is possible to measure respiration by intact tissues it is often useful to be able to isolate uncontaminated, intact and functional mitochondria. Here, we will consider some of the principles for isolating cell organelles. This discussion is not meant to be exhaustive nor will it be a cookbook with recipes to be followed step-by-step.

Choice of Material

If the aim of the experiment is to investigate a specific physiological problem, then the plant material is often well-defined. However, in many cases there is some flexibility in the choice of material and it is therefore wise to follow certain rules of thumb:

- Choose a tissue that is easy to obtain in relatively large amounts (at least 100–200 g fresh weight), such as tubers or hypocotyls.
- Choose a tissue for which a good protocol is available. It can take quite a lot of time and effort to optimize an isolation protocol.

Note that the isolation of mitochondria from green tissues presents special problems because of the large amount of chloroplasts. Roots are a notoriously difficult tissue for isolation of mitochondria.

Disruption of the Tissue

The aim is to rupture as many of the cells in the tissue as possible in order to release their cytoplasm with the mitochondria. Whereas many mammalian tissues are easy to homogenize, plant tissues are much tougher mainly because of the cell walls. There are a number of standard homogenizers, including mixers with rotating knives or razor blades, juice extractors, ball mills, and mortar and pestle; the most suitable method for the specific plant material should be chosen. This may involve some initial optimization experiments.

Homogenization Medium

The medium for homogenizing rat liver only contains an osmoticum and a buffer. That will not work with plant tissues because of the disruption of the large central vacuole (absent in mammalian cells), which releases its fairly acidic content often containing phenols and other potentially harmful substances.

Let us consider the components of the optimal homogenization medium for isolating intact and functional organelles from plant material, one by one.

Osmoticum. This is required to prevent the mitochondria from losing their matrix content because of swelling and bursting. For the isolation of plant mitochondria, there is a tradition of using sucrose or mannitol at 0.2–0.5 M; whereas 0.15 M KCl is standard for the isolation of mammalian mitochondria. In neither case does this mimic precisely the conditions in the cytosol, but it works.

Buffer. The pH in the cytosol of plant cells is around 7.5, and the homogenate should be buffered at pH 7–8. A buffer should therefore be included and a plethora of different buffers are available. Choose an inexpensive one with a pKa of 7–8 (e.g., 4-morpholinopropane sulfonic acid = MOPS) and be sure to use enough to counterbalance the acidic vacuolar content. The tissue type, the tissue/medium ratio, and the buffer concentration determine whether the acidic vacuolar content is properly neutralized. Obviously, special tissues like lemons or CAM plants harvested late at night or in the early morning, require special treatment!

Metal ion chelators. EDTA is usually included to bind mainly Ca^{2+} and Mg^{2+}, which may otherwise activate proteolytic enzymes.

Antioxidants. Antioxidants (reductants) such as ascorbic acid, dithiothreitol, or cysteine are used mainly to protect sulfhydryl groups on proteins against oxidation.

Bovine serum albumin (BSA). This protein from the blood of cows is usually included to protect the mitochondria: BSA binds fatty acids (its function in the blood) released by the action of phospholipases on phospholipids. BSA may also bind phenolics that are released from the vacuole. Both of these groups of compounds can otherwise be troublesome later. In addition, BSA may prevent proteases from degrading mitochondrial proteins by serving as an alternative substrate present at high concentrations.

Polyvinylpyrrolidone (PVP). A polymer that is often included, especially when using leaf material, to adsorb phenolic compounds.

Filtration

To remove larger tissue fragments, cell walls, and often starch grains, the homogenate is always filtered, usually through a nylon net with a mesh size of 30–100 μm.

Purification of Mitochondria

The first step of the purification is usually differential centrifugation. This means that the filtered homogenate is centrifuged first at low speed for a short time (e.g., 3000 g for 5–10 min.) and the pellet (containing cell wall fragments, starch grains, nuclei, intact plastids) is discarded. The supernatant is then centrifuged faster and for a longer time (e.g., 10,000 g for 10–20 min.) and the pellet (termed *crude mitochondria*) is collected. It contains intact and damaged mitochondria as well as peroxisomes and plastid membranes. The supernatant is discarded.

As soon as the mitochondria have been pelleted and thus removed from potentially damaging soluble substances in the homogenate, the medium can be greatly simplified. Usually the so-called wash medium contains only osmoticum and weak buffer (no need for a strong buffer since there is no vacuole content to neutralize). However, it may be wise to include EDTA to bind divalent cations as well as BSA to adsorb any fatty acids released by the action of phospholipases. The resuspended crude mitochondria are often diluted to 50–100 ml with wash medium and subjected to another round of slow and fast centrifugation to remove more contaminants.

The aim of the next step is to separate the intact mitochondria from the contaminants. This usually involves a density gradient centrifugation with Percoll either alone (e.g., Struglics et al. 1993) or together with other dense materials (e.g., Day et al. 1985). Percoll (colloidal silica coated with PVP) is inert and osmotically inactive. It has the further advantage that a Percoll gradient is self-generating because the Percoll microspheres pellet very slowly during the centrifugation. The crude mitochondria are layered on top of the wash medium containing the appropriate concentration of Percoll (usually 25–30%). If the Percoll concentration and the speed and duration of the centrifugation have been chosen correctly, the intact mitochondria will form a distinct band, well away from the contaminants. This band can be collected, washed free of Percoll by repeated dilution and centrifugation, and either used immediately, or frozen with 5% (v/v) dimethylsulfoxide (to preserve integrity) in liquid nitrogen and stored at -80°C until use. This preparation can be stable for months, retaining the ability to form a proton electrochemical gradient over the inner membrane.

The yield is 1–50 mg mitochondrial protein per 100 g fresh weight, depending on the tissue. This represents only a small fraction of the total amount of mitochondria in the tissue. The purification of mitochondria from green leaves presents special problems and a good protocol is described in Day et al. (1985). There is, at present, no good protocol for the isolation of both pure and functional mitochondria from Arabidopsis leaves (Hausmann et al. 2003; Keech et al. 2005), but a protocol for the purification of mitochondria from Arabidopsis cell cultures is given in Kruft et al. (2001). Finally, Boutry et al. (1984) demonstrate how purified mitochondria can be isolated from small amounts of plant tissue (down to 0.5 g fresh weight).

Mitochondria are rod-shaped organelles that can be considered the power generators of the cell, converting oxygen and nutrients into adenosine triphosphate (ATP). ATP is the chemical energy "currency" of the cell that powers the cell's metabolic activities.

This process is called aerobic respiration and is the reason animals breathe oxygen. Without mitochondria (singular, mitochondrion), higher animals would likely not exist because their cells would only be able to obtain energy from anaerobic respiration (in the absence of oxygen), a process much less efficient than aerobic respiration. In fact, mitochondria enable cells to produce 15 times more ATP than they could otherwise, and complex animals, like humans, need large amounts of energy in order to survive.

The number of mitochondria present in a cell depends upon the metabolic requirements of that cell, and may range from a single large mitochondrion to thousands of the organelles. Mitochondria, which are found in nearly all eukaryotes, including plants, animals, fungi, and protists, are large enough to be observed with a light microscope and were first discovered in the 1800s. The name of the organelles was coined to reflect the way they looked to the first scientists to observe them, stemming from the Greek words for "thread" and "granule." For many years after their discovery, mitochondria were commonly believed to transmit hereditary information. It was not until the mid-1950s when a method for isolating the organelles intact was developed that the modern understanding of mitochondrial function was worked out.

The elaborate structure of a mitochondrion is very important to the functioning of the. Two specialized membranes encircle each mitochondrion present in a cell, dividing the organelle into a narrow intermembrane space and a much larger internal matrix, each of which contains highly specialized proteins. The outer membrane of a mitochondrion contains many channels formed by the protein porin and acts like a sieve, filtering out molecules that are too big. Similarly, the inner membrane, which is highly convoluted so that a large number of infoldings called cristae are formed, also allows only certain molecules to pass through it and is much more selective than the outer membrane. To make certain that only those materials essential to the matrix are allowed into it, the inner membrane utilizes a group of transport proteins that will only transport the correct molecules. Together, the various compartments of a mitochondrion are able to work in harmony to generate ATP in a complex multi-step process.

Mitochondria are generally oblong organelles, which range in size between 1 and 10 micrometers in length, and occur in numbers that directly correlate with the cell's level of metabolic activity. The organelles are quite flexible, however, and time-lapse studies of living cells have demonstrated that mitochondria change shape rapidly and move about in the cell almost constantly. Movements of the organelles appear to be linked in some way to the microtubules present in the cell, and are probably transported along the network with motor proteins. Consequently, mitochondria may be organized into lengthy travelling chains, packed tightly into relatively stable groups, or appear in many other formations based upon the particular needs of the cell and the characteristics of its microtubular network.

The mitochondrion is different from most other organelles because it has its own circular DNA (similar to the DNA of prokaryotes) and reproduces independently of the cell in which it is found; an apparent case of endosymbiosis. Scientists hypothesize that millions of years ago small, free-living prokaryotes were engulfed, but not consumed, by larger prokaryotes, perhaps because they were able to resist the digestive enzymes of the host organism. The two organisms developed a symbiotic relationship over time, the larger organism providing the smaller with ample nutrients and the smaller organism providing ATP molecules to the larger one. Eventually, according to this view, the larger organism developed into the eukaryotic cell and the smaller organism into the mitochondrion.

Mitochondrial DNA is localized to the matrix, which also contains a host of enzymes, as well as ribosomes for protein synthesis. Many of the critical metabolic steps of cellular respiration are catalysed by enzymes that are able to diffuse through the mitochondrial matrix. The other proteins involved in respiration, including the enzyme that generates ATP, are embedded within the mitochondrial inner membrane. Infolding of the cristae dramatically increases the surface area available for hosting the enzymes responsible for cellular respiration.

Methods for Establishing Purity, Intactness, and Functionality of the Fractions

It is often important to determine the degree of purity of the mitochondria and their subfractions.

The intactness of the mitochondria is usually established with enzyme assays that take advantage of the fact that an intact membrane is not permeable to certain molecules.

For instance, the outer mitochondrial membrane is impermeable to cytochrome *c*, a 12.5 kDa protein that is too large to pass through the outer membrane pores, and NADH cannot pass through an intact inner membrane.

Functionality is usually tested using an oxygen electrode. If the mitochondria show high rates of oxygen consumption, high respiratory control ratios (>3), and ADP:O ratios close to the theoretical values they are in very good shape. Note that, for mitochondria with an appreciable alternative oxidase activity (measured in the presence of the cytochrome oxidase inhibitor, KCN), the respiratory control ratio and ADP:O ratio should be measured in the presence of an alternative oxidase inhibitor such as salicylhydroxamic acid.

Mitochondrial Activities can be Determined in Situ

It is relatively straightforward to measure the activity of enzymes in the soluble matrix fraction isolated from mitochondria.

However, this fraction is very diluted compared to the matrix inside intact mitochondria (300–500 mg protein/ml!) and protein–

protein interactions may have been lost. When measuring the activity of enzymes in fractions containing membranes—intact mitochondria and submitochondrial particles—a detergent can often be used to remove the permeability barrier that the inner membrane constitutes.

However, it is also possible to use the ionophore alamethicin to permeabilize the inner membrane to molecules <1000 Da, which makes it possible to measure the activity of matrix and inner membrane enzymes in their native proteinaceous environment.

Using alamethicin, it is even possible to measure the activity of cytosolic and mitochondrial enzymes inside intact cells since the ionophore also permeabilizes the plasma membrane.

The Q-Cycle Explains How Protons Can Be Pumped across the Inner Mitochondrial Membrane

The electron transport chain (ETC) in the mitochondrial inner membrane of most eukaryotes consists of four complexes, complex I to complex IV. Complexes I, III and IV pump protons over the inner mitochondrial membrane when transporting electrons.

For each electron pair (corresponding to one NADH or one oxygen atom), four protons are pumped by complex I, four by complex III, and two by complex IV. The detailed three-dimensional structures have been determined for the mammalian complexes III and IV, but not for complex I. The structural information has improved the understanding of the electron paths of complexes III and IV, but the proton-pumping function of the latter is still not fully understood. For complex I, neither electron nor proton paths are known in any detail.

Ubiquinone (Q), also called coenzyme Q, is a mobile electron and proton carrier in the inner mitochondrial membrane, structurally and functionally similar to plastoquinone in the thylakoid membrane of plastids (Figure 1). The ability of Q to carry both electrons and protons—and that it can accept and donate them step-wise—is utilized in electron transport and proton pumping (Figure 2) (Nicholls and Ferguson 2002).

To be reduced, it can receive two electrons from complex I or succinate dehydrogenase and take up two protons from the matrix, forming QH_2. However, at the oxidizing side (complex III), there is actually both oxidation and reduction taking place. A QH_2 will be oxidized to Q at the outer Q-site, releasing two electrons. One of these electrons goes to cytochrome *c*, whereas the other is recycled to reduce an oxidized Q in the inner Q-site—one step to becoming a semiquinone.

The next time a QH_2 is oxidized at the outer Q-site, the semiquinone in the inner site will receive a second electron, becoming a QH_2. So, from two QH_2, we have made one Q and one QH_2, which again can be oxidized. In each cycle (i.e., for each electron passing to cytochrome c), two protons are taken up from the matrix and two protons released to the intermembrane space. So, the Q-cycle mediates the pumping of an extra proton per electron transported, increasing energy conservation.

If we look at the whole reaction from succinate to cytochrome c, we see that for each succinate, four protons are transported from the matrix to the intermembrane space.

However, at the same time, the two electrons are also transported out. So, this corresponds to the net pumping out of two positively charged species. Since, energetically, the mitochondrial membrane potential (which determines the energetic cost of transporting a positive charge out) is much larger than the pH gradient (which determines the cost for electroneutral export of a proton), the complex III proton pumping is less energy-demanding than, for example, complex I, which pumps four protons in their completeness.

Multiple Energy-Conservation Bypasses in Oxidative Phosphorylation of Plant Mitochondria

The electron transport chain (ETC) in the mitochondrial inner membrane of most eukaryotes consists of four large protein complexes, complex I to complex IV. The electron transport activity of complexes I, III and IV (the latter two, often jointly called the cytochrome pathway) is coupled to extrusion of protons across the inner mitochondria.

The electrochemical proton gradient that is formed and maintained in the process is used by the ATP synthase (also called complex V) to make ATP (Figure 1). The plant ETC, however, is highly branched. As a consequence, the plant oxidative phosphorylation system has several alternative pathways of electron transport via type II NAD(P)H dehydrogenases and the alternative oxidase (AOX), and additionally, an alternative proton transporter called the uncoupling protein (UCP).

These proteins mediate bypasses around the proton-translocating multi-protein complexes (I–V): type II NAD(P)H dehydrogenases bypass complex I, AOX bypasses complexes III and IV, and UCP bypasses the ATP synthase (Siedow and Umbach 1995; Vanlerberghe and McIntosh 1997; Møller 2001; Vercesi 2001; Millenaar and Lambers 2003; Fernie et al. 2004; Rasmusson et al. 2004).

The type II NAD(P)H dehydrogenases and the AOX transport electrons without pumping protons, whereas the UCP allows proton flow from the intermembrane space to the matrix without ATP synthesis. The consequence of this is that flux through the energy conservation bypasses will neither contribute to respiratory ATP production, nor will they be controlled by cellular adenylate status. The special features of these energy bypass proteins and discuss the consequences of their existence to the plant.

NAD(P)H Dehydrogenases in Plant Mitochondria

Except for yeast, the ETC of all eukaryotes—including plants—contains a proton-pumping NADH dehydrogenase, complex I, which is inhibited specifically by rotenone. Additionally, plant, fungal and protist mitochondria contain different setups of rotenone-insensitive enzymes capable of oxidizing NADH and/or NAD(P)H and passing the electrons on to ubiquinone.

These NAD(P)H dehydrogenases are classified as type II, whereas complex I belongs to type I. The inner mitochondrial membrane in plants contains such NAD(P)H dehydrogenases, facing either the matrix or the intermembrane space, thus oxidizing NAD(P)H either from the matrix or the cytoplasm.

The outer membrane of all mitochondria contains NADH-ferricyanide and NADH-cytochrome *b* reductase activities that are insensitive to rotenone and to the complex III inhibitor, antimycin A.

There are Three NAD(P)H Dehydrogenases on the Inner Matrix Surface of the Inner Membrane

Much of the information about the number of matrix-facing NAD(P)H dehydrogenases has come from experiments using inside-out inner membrane vesicles. Such vesicles have the matrix surface facing the medium and the external NAD(P)H dehydrogenases hidden inside the vesicles and unable to interfere with the assays.

Recently, this method has been complemented with a technique where one uses a peptide called alamethicin to make a channel for the substrate through the inner membrane. Experiments with these methods show the presence of three NAD(P)H dehydrogenase activities on the matrix side of the inner membrane:

- Complex I pumps protons, is Ca^{2+}-independent and is inhibited by rotenone and diphenylene iodonium (DPI). Complex I has a high affinity for NADH.

- ND_{in}(NADH) does not pump protons, is Ca^{2+}-independent and DPI insensitive. The enzyme has a tenfold lower affinity for NADH than complex I.

- ND_{in}(NADPH) does not pump protons, is Ca^{2+}-dependent and very sensitive to DPI.

Because of the contrasting affinities of complex I and ND_{in}(NADH) for NADH, only complex I is engaged at low matrix concentrations of NADH. However, when the NADH concentration in the matrix increases (e.g., if the ETC is restricted by lack of ADP in the cell), ND_{in}(NADH) can become engaged. When electrons are transported through the type II NAD(P)H dehydrogenases, only the two proton pumping sites of the cytochrome pathway are involved in the flow of electrons from matrix NADH to oxygen. As a consequence, the theoretical ADP/O yield is 1.5 unless the AOX or the UCP are active, in which case it will be even lower.

The External NAD(P)H Dehydrogenases are Ca^{2+}-Dependent

Plant mitochondria can oxidize cytosolic NADH and NAD(P)H directly via external dehydrogenases. The NADH and NAD(P)H dehydrogenase activities are catalyzed by separate enzymes. None of the external dehydrogenases pumps protons, which means that, at most, two sites of proton-pumping are involved when the electrons move from the dehydrogenases to oxygen. So, as for ND_{in}(NADH), the theoretical maximum ADP/O yield is 1.5.

The oxidation of exogenous NADH or NAD(P)H by most plant mitochondria requires Ca^{2+}, and as a result, is inhibited by EDTA and EGTA, chelators of di-and trivalent cations. *In vitro*, the Ca^{2+}concentration needed for half-maximal activity of NADH oxidation is 0.1–0.5 μM, depending on ionic environment and respiratory state This is in the same range as the 0.1–0.2 μM free Ca^{2+} thought to be present in the cytosol of a resting plant cell. Therefore, the external NAD(P)H oxidation may become activated by the increased Ca^{2+}-concentration associated with cellular excitation due to external conditions.

Mammalian mitochondria lack external NAD(P)H dehydrogenases. Instead, they have a Ca^{2+}-dependent glycerol-3-phosphate dehydrogenase on the outer surface of the inner membrane, which donates electrons directly to the ubiquinone pool.

The dihydroxyacetone phosphate formed is converted back into glycerol-3-phosphate by a cytosolic NAD-dependent glycerol-3-phosphate dehydrogenase. In this way, the two low-molecular-weight compounds, glycerol-3-phosphate and dihydroxyacetone phosphate, shuttle redox equivalents from the cytosol, across the outer membrane and to the inner membrane to yield the same amount of ATP as is produced by the plant mechanism. Yeast have both external NADH dehydrogenases and a glycerol-3-phosphate shuttle, and since a mitochondrial glycerol-3-phosphate dehydrogenase has been found in plant mitochondria, plants may also have both systems.

Another way to oxidize cytosolic NAD(P)H that is found in mammalian mitochondria is to use the malate/aspartate shuttle to tranfer reductant (a collective term for high-energy electrons carried by, for example, NAD(P)H, ferredoxin or an organic acid redox couple like malate/oxaloacetate) into the matrix as NADH that can be oxidized by the internal NAD(P)H dehydrogenases. However, it is not clear whether this shuttle can work for import of cytosolic reductant into plant mitochondria.

In the opposite orientation, NADH is shuttled from the matrix to the cytosol via the malate/oxaloacetate shuttle. In this way, the external NADH dehydrogenase can also work as a bypass around complex I. Support for this bypass has been seen in the form of an upregulation of external NADH dehydrogenase activity in mutants where complex I or an internal type II NAD(P)H dehydrogenase has been inactivated.

The Molecular Identities of the Rotenone-Insensitive NAD(P)H Dehydrogenases

The type II NADH dehydrogenases from *E. coli* and yeast are small, single polypeptide proteins of around 50 kDa. They oxidize only NADH and contain FAD as an electron-accepting prosthetic group.

Attempts to purify the type II NAD(P)H dehydrogenases have been unsuccessful in identifying the plant proteins. However, the characterization of potato homologues to the type II NADH dehydrogenases in *E. coli* and yeasts have provided candidate plant proteins to try and assign to the different reported activities.

One protein, NDA1, is internal and the other, NDB1, external of the inner mitochondrial membrane. The latter was shown to be an external calcium-dependent NAD(P)H dehydrogenase. Like a related fungal NAD(P)H dehydrogenase, NDB1 contains a motif for Ca-binding, indicating that in the activation of plant and fungal NAD(P)H dehydrogenase by Ca^{2+}, the ion binds directly to the enzyme.

In Arabidopsis, a weed for which the nuclear genome sequence is known, small families of both NDA-type and NDB-type proteins are encoded by nuclear genes. One of the NDA-type proteins has been shown to be an internal NAD(P)H dehydrogenase. Additionally, a third type of type II NAD(P)H dehydrogenase was found in Arabidopsis, the NDC1. This protein is most closely related to cyanobacterial enzymes, suggesting that the gene entered the eukaryotic domain with the plastid, later to be transferred to the nucleus.

The Alternative Oxidase

The presence of the AOX is one of the features that sets plant (and fungal) mitochondria apart from mammalian mitochondria. This enzyme has therefore received a lot of attention from plant scientists and we now have a fairly good understanding of the regulation of its structure and activity.

The AOX is a quinol-oxygen oxidoreductase and it does not pump protons. The oxygenase transfers electrons from ubiquinol (the reduced form of ubiquinone) to oxygen and generates water as the end-product of the reaction, so four electrons must be transferred to oxygen.

The precise mechanism is not known, but the protein has been shown to contain a catalytical di-iron centre. The functional form of the enzyme is a dimer with the two polypeptides either covalently or non-covalently bound to each other (Figure 2). The predicted structure places the protein embedded into the inner leaflet of the inner mitochondrial membrane, with the active site and the regulatory cysteines exposed to the matrix.

When electrons from NADH oxidation flow through the AOX, at least two of the three sites of proton translocation are bypassed, so energy conservation in the form of ATP is much smaller when the AOX is active. The activity of the AOX must therefore be carefully regulated. There are often several AOX isoforms in each plant species— for example, five genes encoding AOX proteins are present in Arabidopsis. The relative amounts of three isoforms of AOX vary with development, and cellular redox status, as well as external factors.

The AOX is present in the inner membrane as a dimer in which the two subunits can be covalently linked through a disulfide bridge between cysteine residues. The reduced non-covalently linked dimer is the active form, and this form appears to be the predominant one *in vivo*. It was recently shown that a mitochondrial thioredoxin system regulates AOX. Thioredoxins regulate enzymes by reducing disulfide bridges, and it is likely that the regulation involves the redox-active cysteines.

The presence of thioredoxin regulation connects the AOX activation level to the mitochondrial NAD(P)H pool and the oxidation of citric acid cycle metabolites by NADP-utilizing enzymes.

The activity of AOX is also stimulated by 2-oxo organic acids—in particular, pyruvate, which activates by interacting with the same cysteine that is involved in dimerization. This activation may be a feed-forward mechanism that upregulates the capacity of the ETC when the glycolytic end-product is available. Both the cytochrome pathway and the AOX pathway use ubiquinol as the substrate. In the absence of pyruvate, the cytochrome pathway is saturated at 40% Q_{red}/Q_{tot} (40% of the UQ is reduced), at which point the AOX pathway is not engaged at all. The AOX pathway starts accepting electrons at 50% reduction and is not fully saturated at 80%.

In the presence of pyruvate, the AOX becomes engaged at a much lower Q_{red}/Q_{tot} (20%), whereas the cytochrome pathway is unaffected. It is puzzling, however, that the normal *in vivo* concentration of pyruvate appears to be higher than the concentration of pyruvate required to give maximal stimulation of AOX activity *in vitro*. It thus appears that there is little regulation of AOX activity via pyruvate under physiological conditions.

AOX Activity Can Be Measured In Vivo

For many years, benzhydroxamic acid derivatives,, were used to quantify the capacity and activity of the AOX pathway in both isolated mitochondria and intact tissues. AOX capacity was determined from the rate of respiration in the presence of KCN (which blocks the cytochrome pathway), and corrected for residual respiration in the presence of both KCN and SHAM.

AOX *in vivo* activity was determined from the decrease in the rate of respiration when SHAM was added in the absence of KCN. The assumption was that the cytochrome pathway was always used first and that the AOX would be engaged only when the cytochrome pathway was saturated. However, as we have seen above, this assumption is not correct. The consequence of this is that *in vivo* activity of AOX cannot be measured using inhibitors alone.

A better method for measuring AOX *in vivo* activity is now available. It makes use of the fact that cytochrome *c* oxidase and AOX discriminates differently between ^{18}O relative to ^{16}O.

The so-called oxygen discrimination of the sample (tissue or mitochondria) can be measured by mass spectrometry: (1) with KCN to resolve the oxygen discrimination by AOX; (2) with SHAM to resolve the oxygen discrimination by cytochrome *c* oxidase; and (3) without inhibitors to resolve the discrimination value resulting from the activity of both enzymes.

From these values, the percentage contribution by AOX and cytochrome *c* oxidase to the total uninhibited rate of respiration can be calculated. An example of the use of the oxygen discrimination technique with an intact tissue is given below. A remaining problem in the estimation of AOX activity *in vivo* is that measurements can only be made in darkness (to avoid interference from photosynthetic oxygen exchange).

Plant Mitochondria also Contain an Uncoupling Protein

Mammalian mitochondria contain neither type II NAD(P)H dehydrogenases nor AOX. However, they have a UCP that increases the proton permeability of the inner mitochondrial membrane and, in that way, dissipates the proton gradient. This is another mechanism for reducing the ATP production and increasing heat production, for example, for thermoregulation.

Surprisingly, plant mitochondria also contain a UCP. We do not know for certain why plant mitochondria should require two different mechanisms for achieving the same end result. The UCP is related to anion carriers. Its *in vitro* transport activity is dependent on free fatty acids and is inhibited by nucleotides like ATP and GTP but stimulated by superoxide. The latter result indicates that the UCP may protect against oxidative stress.

The Physiological Impact of Energy Bypass Proteins

The respiratory rates of some thermogenic flowers are the highest among plants and, in fact, exceed even those of warm-blooded animals. The external NADH dehydrogenase, AOX and UCP appear particularly active in thermogenic flowers where they are primarily responsible for heat production. At the same time it is clear that the respiratory rate of most plant tissues is insufficient to generate an appreciable increase in tissue temperature.

AOX is up-regulated by cold stress, although this is clearly not to heat the tissue. However, during cold stress, reactive oxygen species (ROS) are formed, and also other stress treatments that lead to elevated ROS will induce AOX. ROS (e.g., superoxide and hydrogen peroxide) are formed as by-products of electron transport under aerobic conditions.

ROS can cause damage to proteins, lipids, and DNA and the cell must therefore limit their formation. AOX helps prevent overreduction of the ETC and thus lowers ROS production. Supporting this, Maxwell et al. (1999) demonstrated the importance of the AOX for limiting ROS formation in living plant cells.

Overexpression of AOX lowered the steady state level of ROS, whereas suppression increased it. This result also points to the mitochondrion as an important site of ROS generation in plant cells. With the recent discovery of ROS stimulation of plant UCP and the increased ROS resistance in plants overexpressing UCP, the picture has become more complex. By decreasing the resistance for electron transport through the proton-pumping complexes, UCP activity may, like AOX, lead to decreased overreduction and ROS formation. Another factor speaking in favour of UCP having a role in ROS protection is that the AOX is inhibited by an ROS product.

Both type II NAD(P)H dehydrogenases and AOX may act as reductant overflow enzymes by reoxidizing surplus NADH to oxygen. This can prevent inhibition of the citric acid cycle under conditions in which the ADP concentration is low, and thus the proton-pumping complexes are restricted by a high membrane potential. In this way, the citric acid cycle can supply carbon skeletons for biosynthesis, even if the ATP level in the cell is high.

The concept of overflow is also relevant for photorespiratory conditions, under which there is a massive flow of glycine that has to be oxidized by the mitochondria. As much as half of the NADH produced in this process is probably exported, but the other half has to be oxidized promptly. There are several reasons to believe that a reductant overflow via energy bypass proteins is involved.

The ND_{in}(NADH) and AOX capacity in leaf mitochondria is higher than in mitochondria from heterotrophic tissue, and both proteins are down-regulated by dark treatment. Consistent with these observations, AOX activity *in vivo* is up-regulated by light (Ribas-Carbo et al. 2000), AOX capacity is induced by greening of etiolated tissue, and expression of the *nda*1 gene [encoding an internal NAD(P)H dehydrogenase] in potato and Arabidopsis is light-dependent.

Mutant tobacco plants, lacking the proton-pumping complex I can still survive, apparently aided by the action of ND_{in}(NADH), which oxidizes matrix NADH with a lower ATP yield.

These plants, however, show decreased photosynthetic rates, indicating that the ND_{in}(NADH) cannot fully compensate for the missing complex I. Also, external NAD(P)H dehydrogenases have been suggested to support photosynthesis. These enzymes may reoxidise NAD(P)H that is shuttled out of the chloroplast in order to balance the NAD(P)H to ATP ratio in the stroma. In summary, there are many indications that mitochondria—especially the energy bypass proteins— have roles in supporting photosynthesis, for example by oxidizing excess NAD(P)H.

The AOX pathway is involved in the acidification/deacidification cycle of CAM plants. CAM plants fix CO_2 in the form of malate during the night. This is done by cytosolic PEP carboxylase, and the malate produced is stored in the vacuole.

The CO_2 fixed is released during the daytime by malic enzymes, located in cytosol and mitochondria. Robinson et al. (1992) used the oxygen discrimination technique to measure the contribution of the AOX pathway to whole-leaf respiration in the CAM-plant, *Kalanchoë daigremontiana.*

They found that the respiratory rate of the leaves increased by 35% in the early hours of the day. This increase was due to increased AOX activity. A more detailed time course showed that the AOX activity increased two-to threefold during the early day at the same time as the malate was metabolized. As soon as the malate had disappeared by mid-day, the AOX activity returned to the low level observed during the night. A reasonable conclusion is that a substantial amount of the malate from the vacuole is oxidized via malic enzyme in the mitochondria. This generates NADH in the matrix, which is oxidized to a large extent via the AOX. The released CO_2 can then be refixed by the Calvin cycle in the chloroplast.

F_oF_1-ATP Synthases: The Worlds Smallest Rotary Motors

F_oF_1-ATP synthases (also called F-ATPases) are present in the inner membranes of mitochondria, chloroplasts, and bacteria. These large, multisubunit enzymes consist of a water-soluble catalytic complex (F_1) attached to an integral membrane protein complex (F_0) that transports protons across the membrane. The F_1 complex is composed of at least five different types of subunits (three α, three β, one γ, one δ, and one å).

When the F_1 complex is dissociated from the membrane, it is active as an ATPase. In fact, under the appropriate conditions the intact F_0F_1-ATP synthase can run in reverse and act as a proton pump, using the energy of ATP hydrolysis to move H^+ across the membrane. Our understanding of how ATP is synthesized was advanced by Paul Boyer at the University of California, Los Angeles, who proposed the binding-change mechanism for catalysis by F-ATPases. From the detailed understanding of how the ATPases function, he proposed that the binding-change mechanism for ATP synthesis contains three important components:

1. The major energy-requiring step is not the synthesis of ATP from ADP and P_i, but the release of ATP from the enzyme.

2. Substrate is bound and products are released at three separate but interacting catalytic sites, corresponding to the three catalytic subunits (b subunits). Each catalytic site can exist in one of three conformations: tight, loose, or open.

3. The binding changes are coupled to proton transport by rotation of the g subunit. That is, the flow of protons down their electrochemical gradient through the F_0 complex causes the g subunit to rotate.

Rotation of the γ subunit then brings about the conformational changes in the catalytic complex that allow the release of ATP from the enzyme, and the reaction is driven forward. The reverse occurs when the enzyme functions as an ATP-driven proton pump.

The first two predictions of the binding-change model are supported by many lines of evidence, mainly kinetic studies, and are now generally accepted. However, the prediction of a rotary mechanism for coupling proton flow to ATP synthesis has been more difficult to demonstrate. Two major breakthroughs have led to confirmation of the third prediction, as well.

The first breakthrough was the determination of the crystal structure of the F_1-part of bovine mitochondrial ATPase by the laboratory of John Walker in Cambridge, England. The crystal structure showed that the three catalytic β subunits differ in their conformations and in the nucleotide bound to them, consistent with the binding-change mechanism.

Even more exciting was the discovery that the γ subunit is inserted like a shaft through the centre of the catalytic complex, which consists of three α subunits and three β subunits arranged alternately in a doughnutlike structure. Moreover, the interface between the γ subunit and the α and β subunits is highly hydrophobic.

The hydrophobicity of the interface minimizes the interactions between the subunits, consistent with the rotation of the γ subunit within the hole formed by the catalytic complex. In other words, the γ subunit looks like a molecular bearing lubricated by a hydrophobic interface. Although many questions were answered by elucidation of the crystal structure of the F_1-ATP synthase, the rotational model cannot be tested by a static "snapshot" of the enzyme. Definitive demonstration of rotation requires a video recording of the spinning of the enzyme in real time. But although they are large for proteins, F_1-ATP synthases are still far too small to be visualized in a light microscope.

To visualize the rotation of the enzyme, Masasuke Yoshida and his colleagues at the Tokyo Institute of Technology came up with an ingenious method to make the enzyme much larger than it is. As Figure below shows, they attached an actin filament labelled with a fluorescent dye to the base of the γ subunit using another protein as a "glue." They then attached the F_1 complex upside down to a glass surface.

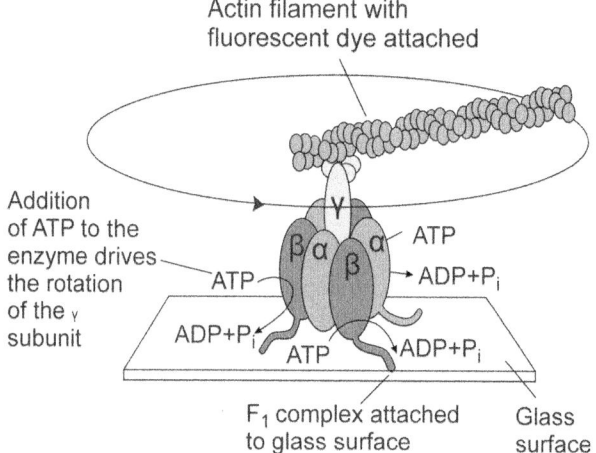

Figure : A method for visualizing rotation of the g subunit. A fluorescently labelled actin filament was attached to one protruding end of the g subunit. The F_1 complex was then attached upside down to a coverslip. When ATP was added to the coverslip, the actin filament rotated. (After Noji et al. 1997.)

If the γ subunit rotates with respect to the catalytic complex, the actin filament should swing around with it. Since the filament is very long compared to the ATP synthase (about 1 μm), its rotation should be visible in a fluorescence microscope. In other words, the fluorescently tagged actin filament, which is large enough to visualize in a light microscope, reports the rotation of the γ subunit.

The results were spectacular! When ATP was added to the modified enzyme, the actin filaments were seen to swing around in a circle at as much as 4 revolutions per second in a fluorescence microscope. To give some idea of scale, if you were a γ subunit, this rotation rate would be equivalent to swinging a several hundred meter-long rod around your head at 4 revolutions per second in water! However, the measured velocity is undoubtedly an underestimation of the actual velocity in vivo, because of the enormous torque required to swing such a large mass. Demonstration of the rotary motion of the γ subunit made it possible to put together a model of how the ATP synthase works. For their contributions to elucidation of the mechanism of ATP synthesis, Paul Boyer and John Walker shared half the Nobel prize in physiology or medicine in 1997. The other half went to Jens Skou for his pioneering work on the K^+,Na^+-ATP synthase, the mammalian counterpart of the plant plasma membrane H^+-ATP synthase.

Transport Into and Out of Plant Mitochondria

Plant mitochondria must exchange metabolites and information (signal transduction) with the cytosol. To give some central examples, the end products of glycolysis, pyruvate and malate, need to be taken up, and citric acid cycle intermediates exported to be used as building blocks in biosynthetic processes.

Likewise, ADP and P_i must be taken up and ATP exported for the mitochondrion to fulfil its role as the powerhouse of the cell. To carry out the transport of this large number of molecules, plant mitochondria possess a family of mitochondrial carrier proteins.

These are transmembrane proteins with a common general structure and they typically transport relatively small charged molecules, like metabolites. In Arabidopsis there are almost 60 members, most of which are not functionally characterized, giving a hint to the complexity of the mitochondrial transport processes. However, mitochondria exchange more than metabolites, and therefore also other types of transporters are involved.

The outer membrane is permeable to molecules of less than 10 kDa, (i.e., all non-polymeric compounds). The permeability results from the presence of pores, which appear to be open in most *in vitro* experiments. Recent studies in animals have characterized pore-forming proteins in the mitochondrial outer membrane, called porins, which appear to regulate transport across the membrane.

Plant mitochondria also have porins but their characterization awaits further research. In our discussion here we will focus on the regulation of mitochondrial transport across the inner mitochondrial membrane. We will first consider the relatively few molecules that can move across the inner membrane without the help of a specific carrier protein. We will then discuss the various metabolite transporters and, finally, briefly look at the transport of proteins and RNA.

Small, Neutral, and Hydrophobic Molecules Diffuse across the Inner Membrane

Irrespective of size, molecules that either are charged or are very hydrophilic can only cross a lipid bilayer with the help of a specific protein carrier.

Small, *neutral* molecules can diffuse across the lipid bilayer and do not require protein transporters. Among those are O_2 and CO_2, two important molecules for respiration. O_2 is needed on the inner surface of the inner membrane where the active sites of both cytochrome c oxidase and the alternative oxidase are located.

The CO_2, produced as an end product in the citric acid cycle, needs to diffuse out of the mitochondrion, ultimately to be lost to the surroundings or re-fixed in the chloroplasts. In addition, however, it is possible that CO_2 can also be transported out of the mitochondrion as bicarbonate (HCO_3^-). Mitochondria are osmotically active and swell in a hypotonic medium. This means that the inner membrane is also permeable to H_2O, but it is not clear whether this occurs by diffusion or if aquaporins may be involved.

For technical reasons, it is of importance that several well-known inhibitors of mitochondrial function can diffuse into mitochondria because they are uncharged and hydrophobic. These are rotenone, the specific inhibitor of complex I, antimycin, the specific inhibitor of complex III, HCN, the protonated form of the cyanide (CN^-) inhibitor of complex IV (and other heme-containing enzymes), and oligomycin, the specific inhibitor of the F_oF_1-ATP synthase.

Transport of Inorganic Ions

Inorganic phosphate ($H_2PO_4^-$) is taken up into the matrix in exchange for a hydroxyl ion (HO^-). The energy for the uptake is supplied by the proton gradient ($-pH$). We will see below that the phosphate gradient, in turn, drives transport of other metabolites.

Several other inorganic ions—Mg^{2+}, Ca^{2+}, K^+, and NH_4^+—are involved in mitochondrial metabolism, but the mechanism by which they pass the inner membrane of plant mitochondria is not well understood. In mammalian mitochondria, uptake of K^+ through a K^+-specific channel, and extrusion via a K^+/H^+ antiporter, is involved in matrix volume regulation and cellular signalling (Garlid and Paucek 2003). A similar system of K^+ transport is present in plant mitochondria.

Both mammalian and plant mitochondria contain an uncoupling protein (UCP). This protein facilitates the movement of protons across the inner membrane and therefore uncouples electron transport and decreases ATP yield.

Transport of Adenylates

One of the important functions of mitochondria is to supply the cell with ATP. To do this, mitochondria have an adenine nucleotide transporter that facilitates the exchange of ADP^{3-} (inward) for ATP^{4-} (outward). This exchange is driven by the membrane potential (inside negative), since there is a net outward movement of one negative charge.

In combination with the phosphate carrier reaction, this means that the uptake of one ADP and one phosphate and the export of one ATP (i.e., the transport event connected to mitochondrial ATP synthesis) costs one vectorial proton—the electrical component for adenylate exchange and the "pH component for phosphate uptake.

Transport of Citric Acid Cycle Intermediates and Related Compounds

Pyruvate and malate, the end products of glycolysis, both have transporters in the inner membrane. Pyruvate is taken up in exchange for a hydroxyl ion, and malate^{2-} in exchange for $H_2PO_4^{2-}$.

The transporter for malate also takes succinate^{2-} and is therefore called the dicarboxylate transporter. Malate^{2-} or succinate^{2-} can, in turn, exchange for citrate^{2-} on the tricarboxylate transporter. Malate^{2-} can also exchange for 2-oxoglutarate^{2-} via a special transporter. An interesting transporter, the oxaloacetate transporter, exchanges oxaloacetate^{2-} for malate^{2-} (or citrate^{2-}), which permits the shuttling of reducing equivalents.

The joint activity of these transporters allows plant mitochondria to import or export most citric acid intermediates. Finally, the glutamate/aspartate transporter facilitates the uptake of glutamate. In mammalian mitochondria, glutamate is taken up together with a proton, and it is thus driven by both the electrical and the chemical proton gradient. A similar transporter is present in plant mitochondria, but a proton co-transport has not been shown, so the transport may be passive in plants. Recently, several additional transporters for metabolites have been identified, and it is likely that certain setups of transporters are only expressed under special developmental conditions, like the mobilization of storage nutrients in developing seedlings.

In several cases the involvement of the mitochondrion in cellular metabolism implies that certain metabolites are taken up across the inner membrane, but the transporter has not yet been described. This is the case for the massive flux of glycine into—and CO_2, serine and NH_4^+ out of—mitochondria in C_3 leaves during photorespiration This is also the case for proline, which accumulates to high concentrations in the cytosol during episodes of plant water stress, and it is then broken down during the rehydration phase by a pathway involving two mitochondrial enzymes.

Transport of Coenzymes

Coenzymes are either synthesized inside the mitochondrion or imported. There is some evidence that, in addition to ATP, at least NAD^+, thiamine pyrophosphate, and coenzyme A are imported across the inner membrane.

The final step in ascorbic acid biosynthesis has recently been shown to be on the outer surface of the inner mitochondrial membrane. In view of the accumulating evidence that ascorbate has an important function in removing ROS inside the mitochondrion, we should expect to find an ascorbate transporter in the inner mitochondrial membrane.

The situation with folate (involved in 1C transfers) is the opposite to that with ascorbate: The only site of synthesis of many of the folate metabolites in the plant cell is the mitochondrial matrix. We can therefore expect that these molecules are exported via specific transporters.

Transport of Reducing Equivalents—Metabolic Shuttles

Mammalian mitochondria cannot oxidize cytosolic NADH or NAD(P)H directly, so they make use of "shuttle mechanisms." An exchange of metabolites by inner membrane transporters takes place through these shuttles so that a reduced compound moves into the matrix in exchange for a more oxidized compound moving out into the cytosol.

In this way, reducing equivalents are transported into the matrix ultimately to be oxidized by the respiratory chain. The shuttles can also be used to transport reducing equivalents in the opposite direction. Although plant mitochondria can oxidize cytosolic NADH and NAD(P)H directly by the two external NAD(P)H dehydrogenases, they can also use these two metabolic shuttles. The malate/oxaloacetate shuttle uses malate dehydrogenase in the cytosol and in the matrix to catalyze the interconversion of malate (reduced) and oxaloacetate (oxidized).

These two compounds are exchanged on the oxaloacetate transporter. This exchange, however, is not driven by the electrochemical proton gradient across the inner membrane, so it can only move reducing equivalents from a relatively reduced compartment to a relatively oxidized compartment. Since the mitochondrial matrix is more reduced than the cytosol, the oxaloacetate transporter is likely to work primarily in the export of reducing equivalents.

Under photorespiratory conditions, large amounts of glycine are oxidized in the mitochondria to give NADH while, at the same time, NADH is required in the peroxisomes to reduce hydroxypyruvate to glycerate. It is thought that as much as 50% of the matrix NADH is exported, probably via the malate/oxaloacetate shuttle.

The more complex malate/aspartate shuttle involves two matrix enzymes and two cytosolic enzymes as well as two inner membrane transporters. In mammals, it is powered by the electrochemical proton gradient in the direction of the import of reducing equivalents. However, since the glutamate/aspartate exchanger may be passive in plants, this shuttle in plants may similarly be independent of the electrochemical proton gradient. If so, it would (like the malate/oxaloacetate shuttle described above) only be able to function as an exporter of mitochondrial reductant.

Transport of Proteins—Protein Import

Plant mitochondria contain their own DNA (mtDNA) and ribosomes and are capable of synthesizing the proteins encoded in the mtDNA. However, plant mtDNA encodes only a few of the proteins found in the mitochondrion. The great majority of the mitochondrial proteins have to be imported from the cytosol where they are synthesized on free ribosomes. These are synthesized with a presequence or targeting peptide in the N-terminal end, which is recognized by receptors on the outer mitochondrial membrane.

The import of proteins requires a complex machinery with two multisubunit protein complexes—one in the outer membrane (TOM, transporter of the outer membrane) and one in the inner membrane (TIM, transporter of the inner membrane). Concomitant with, or immediately afterwards the import of the protein into the correct mitochondrial compartment, the targeting peptide is often cleaved off to yield the mature and active protein.

Transport of RNA

Mammalian mtDNA encodes the full complement of tRNA required for mitochondrial protein synthesis. In contrast, plant mtDNA only encodes some of the tRNAs required for protein synthesis within the mitochondrion. The missing tRNA must be imported by a yet-to-be-elucidated import mechanism.

The Genetic System in Plant Mitochondria has Some Special Features

As in plastids, the genetic system in mitochondria differs both structurally and functionally from the nuclear genome and its gene expression. On the other hand, mitochondria share several genetic features with bacteria. For example, several genes are transcribed together and the ribosomes are of a prokaryotic type. This is in line with the evolutionary origin of the mitochondrion as an endosymbiont of the α-subclass of Proteobacteria, a group including the intracellular pathogen *Ricketsia prowazekii.*

Plants are also distinctly different from other eukaryotes in their mitochondrial DNA (mtDNA) and its expression. For example, the plant mtDNA is very large and highly variable in size between species. The presence of several repeats enables the genome to recombine, changing the gene order or even causing a split into smaller subgenomic molecules, each carrying a subset of the genes. Though the size of mtDNA varies between plants, the number of genes appears to be similar in all plants.

In the fully sequenced 367 kilobasepair (kb) mtDNA of *Arabidopsis thaliana,* the known genes (constituting 65 kb) are scattered in a background of noncoding sequence. A similar situation can be found in the sequenced mtDNA of sugarbeet.

Plant mtDNA is quite different from the much smaller (16 kb) single mtDNA molecule of mammals in which every base pair has a function in coding for protein or mediating transcription and replication, and neither introns nor repeats exist. The expression of the mitochondrial genes in plants is carried out by RNA polymerases that are unlike the nuclear polymerases, but related to the enzymes in bacteriophages.

For several mitochondrial genes, these phage-type polymerases start up transcription from multiple promoters. Many genes in plant mtDNA are broken up into exons by introns. The introns are removed in the processing of the primary RNA into a mature mRNA that can be translated into the correct polypeptide.

The introns in mitochondria belong to the so-called self-splicing type II group, in which splicing is defined by the intron structure. The splicing can also work in the *trans* mode, in which it can join segments from separate RNA molecules, and thus assemble protein-coding reading frames even if the exons are distributed on different primary transcripts. This allows individual exons of a gene to reside independently in the genome, even in different subgenomic DNA molecules.

Because mitochondrial genes are often co-transcribed with unrelated genes into a polycistronic primary transcript, RNA processing also includes endonucleolytic reactions, so that each mature mRNA encodes only one protein. Like mRNAs for nuclear encoded genes, mitochondrial mRNAs are polyadenylated. However, instead of stabilizing (the case for nuclear encoded mRNAs), the mitochondrial mRNA polyadenylation mediates degradation. Instead, stem-loops stabilise the mRNA in plant mitochondria.

In plant mitochondria, an additional RNA processing event takes places—RNA editing. In this process, certain cytidines (C) are deaminated into uridines (U). Rarely, U is aminated to C instead.

This means that the coding sequence is changed at the RNA level, in order to restore the correct information for translation where the DNA has diverged. Editing sites mainly occur in DNA encoding protein sequences and functional RNA, and are rare in noncoding mtDNA. In principle, the higher the importance of an individual base, the higher the probability that it constitutes an editing site. Despite the considerable amount of bases that can be edited (up to 15% of the bases of a gene), the functional relevance of this process and the consequent divergence in mtDNA sequence is not understood. Mitochondrial RNA editing has been found in land plants, but not in algae. It also occurs, though to a lesser extent, in plastids.

The mitochondrial genome encodes only a small proportion of the mitochondrial proteins; less than 40 as compared to up to thousands of nuclear-encoded mitochondrial proteins. Through evolution, genes have been transferred from the mitochondrion to the nucleus. This is a process that still takes place, at least in plants.

For example, subunit 2 of cytochrome oxidase is encoded by the mitochondrial genome in all investigated eukaryotes except for some legumes and protists. Within the legume family, several intermediates of the gene transfer have been found. For example, a functional gene can be found in the mitochondrial genome and an unexpressed one in the nuclear genome, or vise versa.

In order for a mitochondrial gene to be successfully transferred to the nucleus, the protein-coding frame must be integrated into the chromosome, and several other modifications must be made.

For example: (1) a promoter for nuclear transcription must be inserted at the beginning of the gene, (2) the mRNA should acquire sequences specifying the addition of a poly-A tail, and (3) a targeting sequence is necessary for the cytosolically synthesized protein to be imported into the mitochondrion. For proper function, elements regulating the expression in different cell types and under different conditions must also be included. Since the coding sequence for most mitochondrial genes in plants is edited at the RNA level, the mitochondrial genomic sequence will not code for a functional protein if inserted into the nucleus.

Therefore, the DNA that is integrated into the nuclear genome must in most cases be made by reverse transcription from mitochondrial mRNA. Recently, upon sequencing the nuclear genome of *Arabidopsis thaliana*, a copy of most of the mtDNA was found integrated in a nuclear chromosome, indicating that large segments of DNA can move between cellular compartments.

However, most of these mitochondrial genes transferred to the nuclear genome cannot produce functional proteins unless bases corresponding to editing sites are corrected by mutation.

However, this is unlikely to take place before other mutations have further disrupted the sequence. As opposed to other mitochondrial genomes, plant mitochondria utilize the universal genetic code. Thus, this factor will not prevent or delay the transfer of genes between plant organelles as it may do in other eukaryotes.

The special features of the mitochondrial genetic system in plants also have consequences for the study of mitochondrial genes. Due to RNA editing and *trans*-splicing, a reading frame is more difficult to identify based only on genomic sequence. Also, protein sequences in databases are automatically translated from genomic DNA sequences without compensating for RNA editing in many cases. This protocol would not work for mitochondrial genes.

Does Respiration Reduce Crop Yields?

Plant respiration can consume an appreciable amount of the carbon fixed each day during photosynthesis over and above the losses due to photorespiration. To what extent can changes in a plants respiratory metabolism affect crop yields?

Attempts to establish a quantitative relationship between respiratory energy metabolism and the various processes going on in the cell have led to a break-up of respiration into two components. In growth respiration, reduced carbon is processed to bring about the addition of new biomass. The other component, maintenance respiration, is needed to keep existing, mature cells in a viable state. Utilization of energy by maintenance respiration is not well understood, but estimates indicate that it can represent more than 50% of the total respiratory flux.

Although numerous questions remain regarding these issues, there are several empirical examples of relations between plant respiration rates and crop yield. In the forage crop, perennial ryegrass (*Lolium perenne*), yield increases of 10% to 20% were correlated with a 20% decrease in the leaf respiration rate.

Similar correlations have been found for other plants, including corn (*Zea mays*) and tall fescue (*Festaca arundinacea*). However, later investigations have shown that "selection for low rates of mature leaf respiration is not an appropriate method to select for high-yielding cultivars in perennial grasses".

Although there is a potential for increasing crop yield through reduction of respiration rates, a better understanding of the sites and mechanisms that control plant respiration is needed before such changes can be exploited commercially in a systematic fashion by plant physiologists, geneticists, and molecular biologists. Furthermore, much remains to be learned about the general applicability of such observations and the conditions under which slower-respiring lines could be at a disadvantage, causing a reduction in crop yields rather than an increase. Two recent studies illustrate the present difficulty of predicting the effect of directed changes at the molecular level on plant productivity:

Nunes-Nesi et al. (2005) found that the productivity of transgenic tomato plants with a reduced activity of mitochondrial malate dehydrogenase was increased as compared to the wild type. Although the respiratory activity of mitochondria isolated from the transgenic plants was unchanged or higher than in the wild type, the rate of leaf respiration was reduced in the transgenic plants. Photosynthesis was markedly increased in the transgenic lines, possibly linked to increased levels of ascorbate.

Sieger et al. (2005) studied the effect of the alternative oxidase on the growth of tobacco cell cultures. Considering the energy-wasteful nature of the alternative oxidase one might expect that a cell line lacking the enzyme would grow faster than the wild type. Secondly, alternative oxidase has a role in the response of plants to oxidative stress, so a cell line lacking the alternative oxidase would also be expected to handle abiotic stress less well.

The results were surprising—the transgenic cell line with a very low expression of alternative oxidase grew as fast as the wild-type cells under normal, nutrient-sufficient conditions, and faster than the wild type under conditions of macronutrient limitation (low phosphate or low nitrogen).

It appears that the alternative oxidase is an important factor in modulating the growth rate in response to nutrient availability. Plant respiration involves an intricate metabolic network of interacting pathways and a complex regulation of gene expression and enzymatic activities. We clearly need to know much more about these interactions before we can predict the effects of changing the expression of single genes on plant productivity.

The Lipid Composition of Membranes Affects the Cell Biology and Physiology of Plants

The characterization of lipid mutants and the experimental modification of lipid composition by molecular-genetic means have now revealed some broad generalizations about the roles of membrane lipids in the cell biology and physiology of plants.

However, the biochemical and physical bases for the phenotypes that are observed remain uncertain. Changes in fatty acid composition have been shown to alter chloroplast size and architecture. Mutations that eliminate polyunsaturated fatty acids block photosynthesis and prevent the plant from growing autotrophically.

Surprisingly, such mutant plants can grow on sucrose, indicating that photosynthesis is the only plant function that absolutely requires a polyunsaturated membrane. However, polyunsaturated fatty acids are also required as precursors of signalling compounds which are necessary for such diverse functions as pollen development and plant defence, such as jasmonic acid,.

One of the most extensively studied issues in membrane biology is the relationship between lipid composition and the ability of organisms to adjust to temperature changes. For example, chill-sensitive plants experience sharp reductions in growth rate and development at temperatures between 0 and 12°C. The physical and physiological changes in chill-sensitive plants that are induced by exposure to low temperature, together with the subsequent expression of stress symptoms, are termed chilling injury.

Many economically important crops, such as cotton, soybean, maize, rice, and many tropical and subtropical fruits, are classified as chill sensitive. In contrast, most plants of temperate origin are able to grow and develop at chilling temperatures and are classified as chill-resistant plants.

In attempts to link the biochemical and physiological changes associated with chilling injury with a single "trigger" or site of damage, it has been suggested that the primary event of chilling injury is a transition from a liquid-crystalline phase to a gel phase in the cellular membranes. According to this proposal, the transition from liquid-crystalline phase to gel phase would result in alterations in the metabolism of chilled cells and lead to injury and death of the chill-sensitive plants. The degree of unsaturation of the fatty acids would determine the temperature at which such damage occurs. A similar hypothesis has been proposed for chloroplast membranes in which the levels of disaturated phosphatidylglycerol (molecules containing no *cis* double bonds in the fatty acid chains) would determine the chilling sensitivity of plant species.

However, the most recent research indicates that the relationship of membrane unsaturation to plant temperature responses is more subtle and complex than is suggested in these earlier hypotheses. On the one hand, studies of five different *Arabidopsis* mutants have demonstrated that reduced unsaturation can result in plants that grow well at 22°C but are less robust than wild-type plants when grown at 2 to 5°C.

These results were observed even though the lipid changes in most of the mutants are insufficient to cause a lipid phase transition. In addition, the responses of these mutants to low temperature appear quite distinct from classic chilling sensitivity, suggesting that normal chilling injury may not be related to the level of unsaturation of membrane lipids.

In one particular mutant, *fab1*, saturated forms of phosphatidylglycerol account for 43% of the total leaf phosphatidylglycerol—a higher percentage than is found in many chill-sensitive plants. Nevertheless, the mutant was completely unaffected (when compared with wild-type controls) by a range of low-temperature treatments that quickly led to the death of cucumber and other chill-sensitive plants.

A complementary series of experiments has been carried out in tobacco, which is a chill-sensitive plant. The transgenic expression of exogenous genes in tobacco has been used specifically to decrease the level of saturated phosphatidylglycerol or to bring about a general increase in membrane unsaturation.

In each case, damage caused by chilling was alleviated to some extent. These new findings make it clear that the extent of membrane unsaturation or the presence of particular lipids, such as disaturated phosphatidylglycerol, can affect the responses of plants to low temperature. However, membrane lipid composition is not the major determinant of chilling sensitivity in plants.

Utilization of Oil Reserves in Cotyledons

Although the pathway for the mobilization of triacylglycerols has been best characterized in castor bean, it seems to be similar in the storage tissues of other oilseeds.

However, not all seeds quantitatively convert fat to sugar. In castor bean, the endosperm degenerates after the fat and protein reserves are fully utilized. In many oilseeds—such as sunflower (*Helianthus annuus*), cotton (*Gossipium hirsutum*), and members of the squash family (Cucurbitaceae)—the cotyledons become green and photosynthesize after the food reserves are used up.

In these tissues only part of the stored lipid is converted to exported carbohydrate. Much of the lipid-derived carbon remains in the cotyledons, where it contributes to the synthesis of chloroplasts and other cellular structures. Acetyl-CoA is also used directly for the production of energy through respiration. As the greening process takes place, there is a transition in the peroxisome population of these cells: Some peroxisomes have fewer of the characteristics of glyoxysomes and more of leaf-type peroxisomes.

Such a transition is in keeping with the decreased requirement for the breakdown of stored lipids and the increased need to metabolize the products of photorespiration as the tissue goes from a heterotrophic to a more autotrophic mode of metabolism.

Assimilation of Mineral Nutrients

Development of a Root Nodule

The establishment of symbiosis between legumes and rhizobia involves the activation of genes in both the host and the symbiont and an elaborated exchange of signals. The formation of a root nodule, the specialized organ from a plant host that contains the symbiotic nitrogen fixing rhizobia involves two simultaneous processes: infection and nodule organogenesis. The different stages in the initiation and development of a root nodule are shown in Figure below. A detailed description of the infection process.

A recent study of the early infection stages and nodule development in two *Medicago* species and the symbiont, *Rhisobium meliloti* showed that extensive reorganization of microtubules occurred early in the infection process in the pericycle and the inner cortex where the nodule primordium forms. The changes in microtubule organization were observed by immunolocalization, and fluorescence and laser confocal microscopy.

These findings show that upon infection, the pericycle is activated first, followed by the activation of the inner cortex, where the nodule primordium forms. These changes are followed by a major reorganization of the microtubular cytoskeleton network in outer tissues, associated with root hair activation and curling, the formation of pre-infection threads, and the initiation and the growth of an infection network.

Measurement of Nitrogen Fixation

Besides N_2, the nitrogenase enzyme can reduce other substrates such as cyanide, azide and acetylene.

Direct measurement of N_2 fixation (reduction) requires a mass spectrometer, instrumentation that is not readily available, but the reduction of acetylene to ethylene can easily be measured by gas chromatography. Acetylene reduction involves two electrons, $C_2H_2 + 2e^- + 2 H^+ \rightarrow C_2H_4$, as opposed to the eight electrons required for the reduction of N_2 and $2 H^+$:

$$N_2 + 8e^- + 8H^+ + 16ATP \rightarrow 2NH_3 + H_2 + 16ADP + 16P_i$$

Therefore, reduction of four ethylene molecules correspond to the reduction of one N_2 molecule. The acetylene method has limitations, however: Acetylene blocks gas diffusion into nodules, and rhizosphere microorganisms produce ethylene. For these reasons, the acetylene method is suitable for comparative studies, while mass spectroscopy is used for direct measurements of N_2 reduction, which is required for precise quantification of nitrogen fixation.

The Synthesis of Methionine

Methionine and cysteine are the two sulfur-containing amino acids found in proteins. Methionine is synthesized in plastids from cysteine. In the first step of the pathway, cysteine and O-phosphohomoserine react to form cystathionine via the enzyme cystathionine-γ-synthase. Cystathionine is cleaved into homocysteine, pyruvate, and ammonia by cystathionine-β-lysase. Finally, methionine synthase methylates homocysteine to form methionine (Lea 1997).

9

Plant-water Relations

Water is the most abundant constituent of all physiologically active plant cells. Leaves, for example, have water contents which lie mostly within a range of 55–85% of their fresh weight. Other relatively succulent parts of plants contain approximately the same proportion of water, and even such largely nonliving tissues as wood may be 30–60% water on a fresh-weight basis. The smallest water contents in living parts of plants occur mostly in dormant structures, such as mature seeds and spores. The great bulk of the water in any plant constitutes a unit system. This water is not in a static condition. Rather it is part of a hydrodynamic system, which in terrestrial plants involves absorption of water from the soil, its translocation throughout the plant, and its loss to the environment, principally in the process known as transpiration.

Cellular Water Relations

The typical mature, vacuolate plant cell constitutes a tiny osmotic system, and this idea is central to any concept of cellular water dynamics. Although the cell walls of most living plant cells are quite freely permeable to water and solutes, the cytoplasmic layer that lines the cell wall is more permeable to some substances than to others. If a plant cell in a flaccid condition—one in which the cell sap exerts no pressure against the encompassing cytoplasm and cell wall—is immersed in pure water, inward osmosis of water into the cell sap ensues. This gain of water results in the exertion of a turgor pressure against the protoplasm, which in turn is transmitted to the cell wall. This pressure also prevails throughout the mass of solution within the cell. If the cell wall is elastic, some expansion in the volume of the cell occurs as a result of this pressure, although in many kinds of cells this is relatively small.

If a turgid or partially turgid plant cell is immersed in a solution with a greater osmotic pressure than the cell sap, a gradual shrinkage in the volume of the cell ensues; the amount of shrinkage depends upon the kind of cell and its initial degree of turgidity. When the lower limit of cell wall elasticity is reached and there is continued loss of water from the cell sap, the protoplasmic layer begins to recede from the inner surface of the cell wall. Retreat of the protoplasm from the cell wall often continues until it has shrunk toward the centre of the cell, the space between the protoplasm and the cell wall becoming occupied by the bathing solution. This phenomenon is called plasmolysis.

In some kinds of plant cells movement of water occurs principally by the process of imbibition rather than osmosis. The swelling of dry seeds when immersed in water is a familiar example of this process.

Stomatal Mechanism

Various gases diffuse into and out of physiologically active plants. Those gases of greatest physiological significance are carbon dioxide, oxygen, and water vapour. The great bulk of the gaseous exchanges between a plant and its environment occurs through tiny pores in the epidermis that are called stomates. Although stomates occur on many aerial parts of plants, they are most characteristic of, and occur in greatest abundance in, leaves.

Transpiration Process

The term transpiration is used to designate the process whereby water vapour is lost from plants. Although basically an evaporation process, transpiration is complicated by other physical and physiological conditions prevailing in the plant. Whereas loss of water vapour can occur from any part of the plant which is exposed to the atmosphere, the great bulk of all transpiration occurs from the leaves. There are two kinds of foliar transpiration: (1) stomatal transpiration, in which water vapour loss occurs through the stomates, and (2) cuticular transpiration, which occurs directly from the outside surface of epidermal walls through the cuticle. In most species 90% or more of all foliar transpiration is of the stomatal type.

Transpiration is a necessary consequence of the relation of water to the anatomy of the plant, and especially to the anatomy of the leaves. Terrestrial green plants are dependent upon atmospheric carbon dioxide for their survival. In terrestrial vascular plants the principal carbon dioxide–absorbing surfaces are the moist mesophyll cells walls which bound the intercellular spaces in leaves. Ingress of carbon dioxide into these spaces occurs mostly by diffusion through open stomates. When the stomates are open, outward diffusion of water vapour unavoidably occurs, and such stomatal transpiration accounts for most of the water vapour loss from plants. Although transpiration is thus, in effect, an incidental phenomenon, it frequently has marked indirect effects on other physiological processes which occur in the plant because of its effects on the internal water relations of the plant.

Water Translocation

In terrestrial rooted plants practically all of the water which enters a plant is absorbed from the soil by the roots. The water thus absorbed is translocated to all parts of the plant. The mechanism of the "ascent of sap" (all translocated water contains at least traces of solutes) in plants, especially tall trees, was one of the first processes to excite the interest of plant physiologists.

The upward movement of water in plants occurs in the xylem, which, in the larger roots, trunks, and branches of trees and shrubs, is identical with the wood. In the trunks or larger branches of most kinds of trees, however, sap movement is restricted to a few of the outermost annual layers of wood.

Root pressure is generally considered to be one of the mechanisms of upward transport of water in plants. While it is undoubtedly true that root pressure does account for some upward movement of water in certain species of plants at some seasons, various considerations indicate that it can be only a secondary mechanism of water transport.

Upward translocation of water (actually a very dilute sap) is engendered by an increase in the negativity of water potential in the cells of apical organs of plants. Such increases in the negativity of water potentials occur most commonly in the mesophyll cells of leaves as a result of transpiration.

Water Absorption

The successively smaller branches of the root system of any plant terminate ultimately in the root tips, of which there may be thousands and often millions on a single plant. Most absorption of water occurs in the root tip regions, and especially in the root hair zone. Older portions of most roots become covered with cutinized or suberized layers through which only very limited quantities of water can pass.

Whenever the water potential in the peripheral root cells is less than that of the soil water, movement of water from the soil into the root cells occurs. There is some evidence that, under conditions of marked internal water stress, the tension generated in the xylem ducts will be propagated across the root to the peripheral cells. If this occurs, water potentials of greater negativity could develop in peripheral root cells than would otherwise be possible. The absorption mechanism would operate in fundamentally the same way whether or not the water in the root cells passed into a state of tension. The process just described, often called passive absorption, accounts for most of the absorption of water by terrestrial plants.

The phenomenon of root pressure represents another mechanism of the absorption of water. This mechanism is localized in the roots and is often called active absorption. Water absorption of this type only occurs when the rate of transpiration is low and the soil is relatively moist. Although the xylem sap is a relatively dilute solution, its osmotic pressure is usually great enough to engender a more negative water potential than usually exists in the soil water when the soil is relatively moist. A gradient of water potentials can thus be established, increasing in negativity across the epidermis, cortex, and other root tissues, along which the water can move laterally from the soil to the xylem.

Plant Mineral Nutrition

The relationship between plants and all chemical elements other than carbon, hydrogen, and oxygen in the environment. Plants obtain most of their mineral nutrients by extracting them from solution in the soil or the aquatic environment. Mineral nutrients are so called because most have been derived from the weathering of minerals of the Earth's crust. Nitrogen is exceptional in that little occurs in minerals: the primary source is gaseous nitrogen of the atmosphere.

Some of the mineral nutrients are essential for plant growth; others are toxic, and some absorbed by plants may play no role in metabolism. Many are also essential or toxic for the health and growth of animals using plants as food. Six basic facts have been established: (1) plants do not need any of the solid materials in the soil—they cannot even take them up; (2) plants do not need soil microorganisms; (3) plant roots must have a supply of oxygen; (4) all plants require at least 14 mineral nutrients; (5) all of the essential mineral nutrients may be supplied to plants as simple ions of inorganic salts in solution;

and (6) all of the essential nutrients must be supplied in adequate but nontoxic quantities. These facts provide a conceptually simple definition of and test for an essential mineral nutrient. A mineral nutrient is regarded as essential if, in its absence, a plant cannot complete its life cycle.

Nutrients which plants require in relatively large amounts, that is, the essential macronutrients, are nitrogen, sulfur, phosphorus, calcium, potassium, and magnesium. Iron is not required in large amounts and hence is regarded as an essential micronutrient or trace element. With the progressive development of better techniques for purifying water and salts, the list of essential nutrients for all plants has expanded to include boron, manganese, zinc, copper, molybdenum, and chlorine. Evidence has accumulated in support of nickel being essential. In addition, sodium and silicon have been shown to be essential for some plants, beneficial to some, and possibly of no benefit to others. Cobalt has also been shown to be essential for the growth of legumes when relying upon atmospheric nitrogen. Claims that two other chemical elements (vanadium and selenium) may be essential micronutrients have still to be firmly established.

Mineral nutrients may be toxic to plants either because the specific nutrient interferes with plant metabolism or because its concentration in combination with others in solution is excessive and interferes with the plant's water relations. Other chemical elements in the environment may also be toxic. High concentrations of salts in soil solutions or aquatic environments may depress their water potential to such an extent that plants cannot obtain sufficient water to germinate or grow. Some desert plants growing in saline soils can accumulate salt concentrations of 20–50% dry weight in their leaves without damage, but salt concentrations of only 1–2% can damage the leaves of many species.

A number of elements interfere directly with other aspects of plant metabolism. Sodium is thought to become toxic when it reaches concentrations in the cytoplasm that depress enzyme activity or damage the structure of organelles, while the toxicity of selenium is probably due to its interference in metabolism of amino acids and proteins. The ions of the heavy metals, cobalt, nickel, chromium, manganese, copper, and zinc are particularly toxic in low concentrations, especially when the concentration of calcium in solution is low; increasing calcium increases the plant's tolerance. Aluminium is toxic only in acid soils. Boron may be toxic in soils over a wide pH range, and is a serious problem for sensitive crops in regions where irrigation waters contain excessive boron or where the soils contain unusually high levels of boron. All plants grow poorly on very acid soils (pH d" 3.5); some plants may grow reasonably well on somewhat less acid soils. Several factors may be involved, and their interactions with plant species are complex. The harmful effects of soil acidity in some areas have been exacerbated by industrial emissions resulting in acid rain and in deposition of substances which increase the acidity on further reaction in the soil, with consequent damage to plants and animals in these ecosystems.

The elemental composition of plants is important to the health and productivity of animals which graze them. With the exception of boron, all elements which are essential for plant growth are also essential for herbivorous mammals. Animals also require sodium, iodine, and selenium and, in the case of ruminant herbivores, cobalt. As a result, animals may suffer deficiencies of any one of this latter group of elements when ingesting plants which are quite healthy but contain low concentrations of these elements. In addition, nutrients in forage may be rendered unavailable to animals through a variety of factors that prevent their absorption from the gut. Plants and animals differ also in their tolerance of high levels of nutrients, sometimes with deleterious results for grazing animals. For example, the toxicity of high concentrations of selenium in plants to animals grazing them, known as selenosis, was recognized when the puzzling and long-known "alkali disease" and "blind staggers" in grazing livestock in parts of the Great Plains of North America were shown to be symptoms of chronic and acute selenium toxicity.

Essential Points in the Nutrition of Plants

The Nature of Plant Growth Processes

The processes concerned in the growth of plants are the subjects of study by plant physiologists and plant biochemists. A comprehensive account of these processes is outside the scope of the present work, the special object of which is to deal with the outward and visible signs of imperfections in the plant's activities caused by faulty mineral nutrition. Nevertheless it is useful to have before us the general features of the main processes involved and to realize that the symptoms we shall be discussing later have a physiological basis, and are not direct and unchangeable signs of the specific deficiencies but result from the derangement of the complicated mechanism of the plant's vital activities.

The main processes involved in plant development may be summarized as follows:

Absorption: Intake of water and mineral elements by the root system.

Carbon assimilation or photosynthesis: Intake of carbon dioxide from the air by the leaves, and reaction of the gas with water in the leaf in the presence of the green chlorophyll to form sugar and free oxygen.

Formation of protoplasm: Protoplasm is the living material of the plant, consisting mainly of proteins, complex compounds of nitrogen built up by the plant from more simple compounds of this element.

Respiration: The combination of oxygen with various food substances synthesized by the plant, especially sugars, whereby energy is produced.

Transpiration: Loss of water from the plant, mainly from the leaves.

Translocation: The movement of materials within the plant.

Storage: Storage of reserve products in various organs and tissues.

During growth there are continuous processes of building up of complex compounds of carbon and nitrogen and breaking down of these into more simple substances, in which water and oxygen are intimately concerned. These processes together comprise plant metabolism.

In the course of the metabolic processes innumerable substances are formed, such as sugars, starch, cellulose, acids, lignin, tannins, amino acids, proteins, amides etc., and many plants also produce special products, as for instance nicotine in the tobacco plant.

For the normal functioning of the above processes there must be an adequate intake of water by the plant to maintain the plant cells in a more or less turgid condition and, since water is being continuously lost at a varying rate from the plant, intake and movement within the plant tissues must be capable of ready adjustment to these changes.

As a result of metabolic activities plants develop special organs of growth and reproduction, each of which has its special characters and makes particular demands on the nutrient supplies of the plant.

With all plants there are well defined seasonal growth cycles. Thus annuals, such as cereals, begin from the seed, give rise to seedlings, which later flower, form grain and ripen off, whilst perennial deciduous trees, such as apples, pears, etc., begin growth in the spring, using stored reserves of food, form leaves, make shoots, blossom and form fruits and subsequently shed their leaves, but meanwhile pass on reserve foods to various storage organs in preparation for the next season's growth. Coincident with these growth cycles there are well defined chemical cycles of nutrient elements and elaborated products in the leaves, stems and roots, etc. It will be shown later that these cycles are of great importance in considering p-deficiency effects and in diagnosing their causes.

The Plant Environment

Nutritional problems must be considered in relation to all the conditions in which plants live, and not merely in terms of the amounts of plant nutrients contained in or added to the soil. For example, those who are accustomed to growing plants know that the temperature must not be too low or no growth may result or that if too high the plants may be injured. An optimum temperature is usually recognized and this may vary according as to whether the plant is young or old. Similarly the importance of light is well known and plants may be put in special positions to obtain a maximum supply of light at one stage and may be shaded at another.

The actual duration of the daily period of illumination also affect growth and there are plants which are classified as requiring "long day" conditions to complete their growth cycles and others as needing "short day" conditions. If the special "long" or "short" day periods are not forthcoming for the respective classes of plants requiring these, their growth cycles are abnormal and they may fail entirely to produce flowers, grain or fruit.

The humidity of the atmosphere, as distinct from the water supply in the soil, is of importance in determining the water conditions within the plant, as these are dependent on both water intake by the roots and water loss from the leaves, the latter being largely influenced by the air humidity.

Even the presence of adequate quantities of plant nutrients in the soil is no guarantee that they will be absorbed by the plant roots. It will be shown later how these may be present in forms which are not available to the plants, but even when they would be considered as being present in suitable forms for absorption, other factors may prevent this taking place. An example of this latter condition is afforded in poorly aerated soils where lack of oxygen near the roots may prevent them from actively absorbing mineral nutrients

The problems of such influences in the plant environment as those just mentioned are complicated by the fact that they do not act independently, but their effects are modified by one another. Thus the effects of light intensity or period of daylight may vary with different temperature conditions.

The requirements of plants for different nutrients may be affected by conditions of light, temperature and water supply, and by other factors of the general environment. Thus the need for nitrogen may be less under conditions of relatively low light intensity whereas the need for potash in these circumstances may be greater, these facts being of importance in growing tomatoes under glass. The effect of nitrogen in relation to light may be shown by growing a plant under normal light conditions with insufficient nitrogen, when the leaves will show the well known symptoms of nitrogen deficiency-pale green, yellow, orange and red tints. If such a plant be then shaded, the leaves will turn a darker green and growth may be visibly increased. It can be shown that the lowered light conditions result in an increase of "soluble" a breaking down of proteins, thereby rendering the nitrogen of these available for growth processes.

This interrelationship of environmental factors is well illustrated by an experiment on apple trees at Long Ashton.

Bramley's Seedling trees were grown in compost in large pots and given a small dressing of a nitrogenous fertilizer. Some of the trees were grown in a specially constructed glass house and an equal number in an adjoining wire enclosure. The trees in the enclosure showed severe symptoms of nitrogen deficiency-pale green and yellow leaves, reddish brown barks and highly coloured, red fruits. The condition was corrected by further dressings of nitrogen. In contrast, the trees under glass, where the light was of less intensity and the temperature higher, made vigorous growth, carried large, green leaves and bore large, green fruits.

Iron and zinc deficiency symptoms may be less severe under conditions of low light intensity, whilst boron deficiency effects are less severe and magnesium deficiency effects are more pronounced in wet seasons than in dry ones.

The rate of water absorption is less at lower temperatures than at higher ones and efficient intake is also dependent on good aeration. These facts may result in a water deficit within plants growing in cold, wet soils when the air temperature is high. The raw materials needed for plant growth consist of carbon dioxide, which is obtained from the atmosphere through the stomata of the leaves, and water and the so-called mineral nutrients, which normally enter the plant through the medium of the roots.

The importance of water and carbon dioxide in the nutrition of plants will be apparent from the facts that water often comprises 80 to 90 % of the total weight of growing plants, and carbon and oxygen together may account for over 80% of their dry matter, i.e., the solid matter remaining after water is removed. As against these large amounts, the mineral nutrients, as measured by the ash content of the plants, i.e., the mineral residue obtained when the organic matter is destroyed by heat, often contribute from 5 to 15% of the dry matter.

It has been shown in recent years that certain organic compounds, known as "growth promoting substances" or "hormones", which occur in plants, and some of which are also present in soils and natural manures, are capable of producing marked growth responses, such as increased root growth, shoot and leaf curvatures, stimulation or suppression of buds, increased fruit setting, prevention of fruit abscission etc. They appear to perform important functions in the growth of plants. Examples of substances of this kind which can produce growth responses are 13 indole-acetic acid, 13 indole-butyric acid, phenyl acetic acid, A, naphthalene-acetamide, vitamin B_1

It is not at present clear to what extent growth substances are absorbed by plants from soils, although it has been shown that vitamin B_1, which occurs naturally in soils, can be obtained in this way.

The Mineral Nutrients

It has been shown in numerous researches that certain elements are necessary for the healthy growth of plants. They are sometimes spoken of as essential elements and, since some are needed in relatively large quantities and others in very small amounts, the former are referred to as "major" elements and the latter as "minor" or "trace" elements, or as micro nutrients.

The terms "major" and "minor" do not refer to the relative importance of the functions of the elements in plant growth, and for this reason the term "trace" element is preferable for the latter class. Major elements: Nitrogen, phosphorus, calcium, magnesium, potassium, sulfur.

Trace Elements: Iron, manganese, boron, copper, zinc and molybdenum.

(Iron occupies an intermediate position and is usually included in the major elements group. In dealing with field problems it is more convenient to group it with the trace elements.)

In addition, there are, other elements, such as sodium, chlorine and silicon, which produce beneficial effects on the growth of certain plants but which have not so far been shown to be absolutely essential to growth. The element aluminium is of general occurrence in plants, but seems to be without direct nutritional value, although aluminium sulfate is used, because of its acidifying properties, to change the colour of hydrangeas growing on alkaline soils from pink to blue, and aluminium may also exert indirect influences on nutritional processes.

Other elements often occur in plants but they are not known to serve any useful function and frequently they act as plant poisons or toxins. The nutrient elements can only be absorbed by plants when present in certain forms: nitrogen from nitrates and ammonium salts; phosphorus from phosphates; calcium, magnesium and potassium from their salts; sulfur from sulfates; iron from ferrous or ferric salts (more readily from ferrous salts); manganese from manganous salts; boron from borates; copper and zinc from their salts, and molybdenum from molybdates.

There may appear to be certain exceptions to this statement in practice. For instance, nitrogen may be applied to a soil as "organic" nitrogen, as in hoof meal or urea, and sulfur may be added as the element itself, as in flowers of sulfur, ground sulfur, etc. In such conditions the added materials are, however, converted into the nitrate and sulfate forms respectively by soil organisms before being absorbed by the plants.

Further points of importance in connection with the absorption of the mineral nutrients by plants are as follows:

(a) They must be absorbed from relatively dilute solutions or the plants will be! injured or even killed.

(b) Certain of the elements slow down the absorption of others into the plant, e.g., calcium slows down potassium and vice versa. The phenomenon is known as "antagonism".

(c) Healthy plants result when the nutrients are absorbed in certain relative proportions. When the proportions are suitable the nutrient medium is said to be "balanced". When ratios between nutrients are too wide, deficiency conditions are created. Thus if a high proportion of nitrogen to potassium is absorbed, the plant will suffer from potassium deficiency.

(d) Nutrients, even though present in the nutrient solution in satisfactory amounts and proportions, may not be absorbed by the plant unless the "reaction" of the solution as regards acidity and alkalinity is satisfactory. The reaction is measured in terms of the pH scale, which is merely a convenient notation for stating the conditions of acidity in the solution (strength or intensity of acidity, not total amount). The neutral point is represented by pH 7.0; below this value the solution is acid and above it is alkaline. Many crop plants prefer a reaction slightly on the acid side-pH 6.0 to 6.5 and extreme values are in the neighborhood of 4.0 on the acid side and 9.0 on the alkaline side.

(e) The nutrient medium must contain an adequate supply of oxygen, i.e., aeration must be satisfactory.

The Functions of The Mineral Nutrients

Knowledge of the main functions of the mineral nutrients is useful in helping us to understand the effects produced by deficiencies of any one of them.

Essential Major Elements

Nitrogen. Nitrogen is a major constituent of several of the most important substances, which occur in plants. It is of outstanding importance among the essential elements in that nitrogen compounds comprise from 40 to 50% of the dry matter of protoplasm, the living substance of plant cells. For this reason nitrogen is required in relatively large quantities in connection with all growth processes in plants. It follows directly from this that without an adequate supply of nitrogen appreciable growth cannot take place and that plants must remain stunted and relatively undeveloped when nitrogen is deficient.

Proteins, which are of great importance in many plant organs, e.g., seeds, are compounds of nitrogen whilst chlorophyll, the green colouring matter of the leaves, also contains the element.

From this latter fact it will be apparent that when nitrogen is deficient leaves will contain relatively little chlorophyll, and will thus tend to be pale green in colour. In addition to the above substances, numerous other organic compounds of importance in plants, such as amino acids, amides, and alkaloids, are compounds of nitrogen.

Certain compounds of nitrogen are very mobile in plants, and this enables them readily to mobilize supplies of the element at vital growing points and to transfer stored supplies to points where they are most required. Such transference is common from old tissues to young growing points when supplies of the element are short. This mobility and re-utilization of nitrogen explains why deficiency symptoms of the element always appear first in the older parts of plants and why growing points are the last to be affected.

Phosphorus: This element, like nitrogen, is closely concerned with the vital growth processes in plants as it is a constituent of nucleic acid, and nuclei in which this occurs are essential parts of all living cells. Hence a deficiency of this element will also be expected to result in greatly restricted growth. Phosphorus is also of importance in seeds and in connection with the metabolism of fats. Compounds of phosphorus are concerned with the processes of respiration and with the efficient functioning and utilization of nitrogen. This relationship to nitrogen probably accounts for the fact that several of the symptoms of phosphorus deficiency are identical or similar to those which result from a deficiency of nitrogen, Phosphorus is also of special importance in the processes concerned in root development and the ripening of seeds and fruits.

Calcium: Calcium occurs in plants chiefly in the leaves and the amounts present in seeds and fruits are relatively low. One of its main functions is as a constituent of the cell wall, the middle lamella of which consists largely of calcium pectate. This function appears to be of fundamental importance since, if calcium is replaced by any other of the essential elements, such as magnesium or potassium, the organic materials and mineral salts in the cells are readily leached through the walls. Although a large proportion of the calcium contained in the plant may be soluble in water-as much as 60% in cabbage calcium does not appear to move freely from the older to the younger parts of plants, and hence young tissues contain lower proportions of calcium than older ones. This may explain why calcium deficiency effects begin at the tips of shoots.

Magnesium: The outstanding fact about magnesium is that it is a constituent of chlorophyll, and is essential to the formation of this pigment. As a result, when magnesium is deficient, one of the symptoms commonly shown by plants is chlorosis. Magnesium is also regarded as a carrier of phosphorus in the plant, particularly in connection with the formation of seeds of high oil content, which contain the compound lecithin. The element seems to be very mobile within the plant, and when deficient is apparently transferred from older to younger tissues where it can be re-utilized in the growth processes. This agrees with the observation that signs of magnesium deficiency invariably make their appearance first on the oldest leaves and progress systematically from them towards the youngest ones.

Potassium: Unlike all the other major elements, potassium does not enter into the composition of any of the important plant constituents, such as proteins, chlorophyll, fats and carbohydrates, concerned in plant metabolism. For this reason its role is more difficult to determine, and in spite of much study it cannot be said that the functions of potassium are clearly understood. The element is present in all parts of plants in large or fairly large proportions. It seems to be of special importance in leaves and at growing points, as these are especially rich in potassium. Probably the whole of the potassium in plants is present in soluble form, and most of it seems to be contained in the cell sap and cytoplasm.

It is outstanding among the nutrient elements for its mobility and solubility within the plant tissues, and these properties no doubt account for the ready way in which potassium can be re-utilized by young tissues when the element is in short supply. Among the functions which have been attributed to potassium and the processes with which it may be concerned, the following may be mentioned: The formation of carbohydrates and proteins; the regulation of water conditions within the plant cell and of water loss by transpiration; as a catalyst and condensing agent of complex substances; as an accelerator of enzyme action; as contributing to photosynthesis through its radioactive properties. It has been shown in many instances that the potassium content of plants is frequently much higher than is necessary for healthy growth, and it is generally considered that luxury absorption of potassium often takes place.

The great mobility of potassium in plants, its special importance for and its reutilization by young tissues, and its apparent functions as a regulator of plant processes on a large scale are in harmony with the observations that, when potassium is moderately deficient, the effects are seen first in the older tissues and progress from these towards the growing points, but, when the deficiency is acute, growing points are severely affected, and die-back and general collapse of the plants commonly occur.

Sulfur: Sulfur occurs in plants as a constituent of proteins, and of certain volatile compounds such as mustard oil. It seems to be connected with chlorophyll formation, although it is not a constituent of this substance. Its functions in connection with proteins and chlorophyll doubtless account for the similarity of its deficiency effects to those due to deficiency of nitrogen.

Essential Trace Elements

Iron: Iron is closely concerned with chlorophyll formation but is not a constituent of it. Its role appears in this connection to be that of a catalyst. As a result of this function of iron, chlorosis is invariably an outstanding symptom when the element is deficient. Iron may also act as a catalyst, in the role of an oxygen carrier, in respiration.

A point of great importance in connection with iron is its relative immobility in plant tissues. Its mobility seems to be affected by several factors, such as the presence of manganese, potassium deficiency and high light intensity. There is evidence that the amount of chlorophyll is related to "active" iron in plants.

It will thus be seen that so-called iron deficiency in the plant may in fact usually mean iron immobility. Lack of mobility may also account for the fact that iron deficiency is first shown in the younger tissues.

Manganese: The functions of manganese are regarded as being closely associated with those of iron and as being concerned with chlorophyll formation. Hence, when manganese is deficient, chlorosis is a common symptom. Manganese may decrease the solubility of iron by oxidation and hence, an abundance of manganese within the plant may lead to iron deficiency and chlorosis.

Manganese is regarded as having the functions of a catalyst; its activities being specially concerned with oxidation and reduction reactions within the plant tissues.

Boron: The exact role of boron is not known, but again the evidence as for the other trace elements, suggests that, its functions are those of a catalyst or reaction regulator. It can delay the onset of calcium deficiency effects but cannot replace calcium; it tends to keep calcium soluble; it may act as a regulator of potassium/calcium ratios, and of the absorption of nitrogen; it may be concerned with the oxidation-reduction equilibrium in cells.

Such functions as the above accord with the results which follow from a deficiency of the element, when growth processes show sudden collapse and drastic derangements of metabolism occur.

Zinc and Copper: Although specific functions have not been determined for these elements, here again the evidence points to their roles as catalysts and regulators. Deficiencies of both are associated with chlorosis and a serious general collapse of vital growth processes.

Since catalysts are not used up in the chemical reactions which they promote, we can understand how it comes about that quite small or even minute quantities of the "trace elements", iron, manganese, boron, zinc and copper, may nevertheless be essential to the plant's health and growth.

Elements Sometimes Beneficial

Sodium: As sodium is not strictly an essential element it cannot be expected to have a specific role in the metabolic activities of plants.

Where sodium produces significant effects it is often regarded as a conserver of potassium and as being able partly to replace that element in its role. In no instance, however, has it been shown that sodium can wholly replace potassium where the latter is acutely deficient. In such circumstances, sodium is ineffective as a substitute for potassium, even for sodium-loving plants, such as sugar beet, mangold and barley. Sodium seems to affect the water relations of plants and often enables sugar beet and other crops to withstand drought conditions which would otherwise produce severe adverse effects.

Chlorine: The evidence of the role of chlorine in plants is somewhat contradictory, and no general statement can be made. In tobacco it has been shown to increase the water content of the tissues and to affect carbohydrate metabolism, leading to an accumulation of starch in the leaves. The element is present in plants as chloride, and is wholly soluble.

Bibliography

Alfred Steferud: *Diseases of Fruits and Nuts*, Biotech Books, Delhi, 2005.

Arthur W. Gilbert, Mortier F. Barrus and Daniel Dean: *Growing and Breeding of Potatoes*, Asiatic Pub, Delhi, 2006.

Ashworth S.: *Seed to Seed*, Decorah, Seed Savers Publications, 1991.

Ausubel, F.M. : *Current Protocols in Molecular Biology,* New York: John Wiley and Sons, 1989.

Bahar A. Siddiqui and Samiullah Khan: *Plant Breeding Advances and in vitro Culture*, CBS, Delhi, 1997.

Banki, L.: *Bioassay of Pesticides in the Laboratory,* Akademiai Kiado, Budapest, 1978.

Barbeau, G.: *Tropical Fruits in Nicaragua*, Managua, Nicaragua Ministerio de Desarrollo Agropecuario, Agraria, 1990.

Barnes, N.: *Biology*, New York, Worth Publishers, 1989.

Barnum, Susan R.: *Biotechnology: An Introduction*, Belmont, Thomson/Brooks/ Cole, 2005.

Barrington, E. J. W.: *Biochemistry of Primitive Deuterostomians*, London, Academic Press, 1974.

Baruah, Akhil: *Advanced Morphology of Angiosperms*, Aavishkar, Delhi, 2008.

Batlle I. and Tous J.: *Carob Tree (Ceratonia siliqua L.).* Rome, International Plant Genetic Resources Institute, 1997.

Bauer MW: *Biotechnology-the Making of a Global Controversy*, Cambridge, Cambridge University Press, 2002.

Bhatia, A L : *Biochemistry and Endocrinology,* Indus Valley, Delhi, 2002.

Bhatnagar, Vasudev : *Cell Science and Technology*, Campus Books, Delhi, 2009.

Brandwein, P.F. : *Sourcebook for the Biological Sciences,* San Diego: Harcourt Brace Jovanovich, 1986.

Broach, J.R.: *The Molecular Biology of the Yeast*, Cold Spring Harbor, Cold Spring Harbor Laboratory, 1981.

Brodwin, Paul : *Biotechnology and Culture: Bodies, Anxieties, Ethics*, Bloomington: Indiana University Press, 2001.

Cahill, Lisa : *Genetics, Theology, and Ethics: An Interdisciplinary Conversation*, New York: Crossroad, 2005.

Chaudhary, Vikas: *Entomology and Pest Management*, Navyug, Delhi, 2008.

Chauhan, B.S. : *Principles of Biochemistry and Biophysics*, Laxmi Publications, Delhi, 2008.

Chiranjib Chakraborty: *Advances in Biochemistry and Biotechnology*, Daya, Delhi, 2005.

Chrispeels, Maarten : *Plants, Genes and Crop Biotechnology*, Sudbury MA, Jones and Barlett Publishers, 2003.

Clark, J.M.: *Experimental Biochemistry*, New York, W.H. Freeman and Company, 1977.

Clawson, Calvin: *The Mathematical Traveller*, New York, Plenum Press, 1994.

Collins, Steven : *The Race to Commercialize Biotechnology: Molecules, Market and the State in Japan and the US.*, New York: Routledge, 2004.

Collymore L.: *Fruit Production in Barbados*, Port of Spain, Trinidad and Tobago, 1996.

Coste R.: *Coffee: the Plant and the Product*, London, MacMillan, 1992.

Cronquist, A.: *The Evolution and Classification of Flowering Plants*, New York Botanical Garden, Bronx, New York., 1988

Currah L. and Proctor F. J.: *Onions in Tropical Regions*, Kent, Natural Resources Institute, 1990.

Dabholkar, A.R.: *General Plant Breeding*, Concept, Delhi, 2006.

Daphne C. Elliott: *Biochemistry and Molecular Biology*, Oxford University Press, Delhi, 2005.

David Sadava: *Plants, Genes and Crop Biotechnology*, Sudbury MA, Jones and Barlett Publishers, 2003.

Dixon, Dougal: *After Man-A Zoology of the Future*, New York, St. Martin's Press, 1981.

Dodds, John H.: *Plant Genetic Engineering*, New York, Cambridge University Press, 1985.

Doijode S. D.: *Seed Germination in Fruits*, New Delhi, Malhotra Publishers, 1993.

Duckworth, W. L. H.: *Morphology and Anthropology : A Handbook for Students*, Cosmo, Delhi, 2006.

Dudley, E.: *The Critical Villager: Beyond Community Participation*, London, Routledge, 1993.

E. Ramann: *The Evolution and Classification of Soils*, Asiatic Pub, Calcutta, 2006.

Elliott, B N : *Biochemistry and Molecular Biology*, Oxford University Press, Delhi, 2005.

Featherly H. I.: *Taxonomic Terminology of the Higher Plants*, USA, Iowa State College Press, 1954.

Ferentinos L.: *Proceeding of the Sustainable Taro Culture for the Pacific Conference*, Honolulu, HITAHR, 1993.

Fransman M, Junne G, Roobeek A: *The Biotechnology Revolution?*, Oxford, Blackwell, 1995.

Friedberg, E.C.: *DNA Repair*, New York, WH Freeman and Company, 1985.

Fumento, Michael: *Bioevolution: How Biotechnology is Changing Our World*, San Francisco, Encounter Books, 2003.

Ganguly, Smriti : *Biochemistry of Biomolecules*, Pearl Books, Delhi, 2007.

Ghulam Hassan : *Soil Microbiology and Biochemistry*, New India Publishing Agency, Delhi, 2010.

Goodsell, David S.: *Bionanotechnology: Lessons From Nature*, Hoboken, Wiley-Liss, 2004.

Graf, Alfred Byrd: *Advances in Plant Physiology*, Rajat Pub, Delhi, 2008.

Hardy B.: *Biology and Agronomy of Forage Arachis*, Cali, International Centre for Tropical Agriculture, 1994.

Jeffers P.: *Evaluation of Four Onion Varieties in Montserrat*, Plymouth, CARDI, 1992.

Jones, R. M.: *Plant Resources of South-East Asia*, Wageningen, Pudoc Scientific Publishers, 1992.

Khanna, V K : *Objective Genetics, Biotechnology, Biochemistry and Forestry*, I.K. International Publishing House, Delhi, 2008.

Kuppuram, G & K. Kumudamani: *History of Science and Technology in India*, Delhi, Sundeep Prakashan, 1990.

Kurzweil, Ray: *The Age of Spiritual Machines*, New York, Penguin Books, 1999.

Larry V. McIntire : *Biotechnology: Science, Engineering, and Ethical Challenges for the Twenty-first Century*, Washington, DC: Joseph Henry Press, 1996.

Macself, A.J.: *Soils and Fertilizers*, Satish Serial Pub, Delhi, 2005.

Madan Lal Bagdi: *Physiology, Biochemistry and Biotechnology*, Manglam Pub, Delhi, 2007.

Madulid Domingo A.: *A Pictorial Cyclopedia of Philippine Ornamental Plants*, Philippines, Makati Metro Manila, 1995.

Mahindru, S. N.: *Food Safety and Pesticides*, APH, Delhi, 2009.

Meena Francis: *Biotech's Dictionary of Biochemistry*, Biotech Books, Delhi, 2007.

Muneesh Kainth: *Chordate Embryology*, Dominant, Delhi, 2003.

Nobel, P. S.: *Physicochemical and Environmental Plant Physiology*, Academic Press, San Diego, 1999.

Old, R.W. : *Principles of Gene Manipulation*, London, Blackwell Scientific Publications, 1989.

Oldham P.: *Cost of Production of Major Tree Crops in Dominica*, Roseau, Ministry of Agriculture, 1991.

Parry M.L.: *Climatic Change, Agriculture and Settlements*, Dawson Folkestone UK, 1978.

Paul M. Althouse: *Introduction to Agricultural Biochemistry*, Biotech Books, Delhi, 2005.

Pemberton, R. W.: *Predictable Risk to Native Plants in Weed Biological Control,* Oecologia, 2000.

Qystein V. Sjaastad: *Physiology of Domestic Animals*, International Book Distributing Co., Delhi, 2005.

Ragone D.: *Breadfruit: Artocarpus Altilis (Parkinson) Fosberg*, Rome, International Plant Genetic Resources Institute, 1997.

Rifkin, Jeremy: *The Biotech Century*, New York, Penguin Putnam, 1998

Rutherford Lyn.: *A Gourmet's Book of Mushrooms & Truffles*, Sydney, Golden Press Pvt. Ltd., 1991.

Sharma, Pradeep : *Biochemistry and Organisation of Cells*, RBSA Pub, Delhi, 2006.

Stover, R. H. and Simmonds N. W.: *Bananas*, United Kingdom: Longman Scientific and Technical, 1991.

Swarnim, K. : *A Textbook of Biochemistry and Microbiology*, Surendra Pub, Delhi, 2010.

Tawde, A. B.: *Propagation and Rootstocks of Mango*, New Delhi, Malhotra, 1993.

Urton, Gary: *The Social Life of Numbers*, Austin, University of Texas Press, 1997.

Vanangamudi, K.: *Principles and Methods of Plant Breeding*, International Book, Delhi, 2005.

Whealy K.: *The Garden Seed Inventory*, Decorah, Seed Saver Publications, 1988.

White, G.F.: *Natural Hazards: Local, National, Global*, Oxford University Press, New York, 1974.

Woolfe Jennifer A.: *The Potato in the Human Diet*, Cambridge, Cambridge University Press, 1989.

Yadav, M.: *Nutritional Biochemistry and Metabolism*, Arise Pub, Delhi, 2008.

Index

□□□